FI
With the Purc...

The Ultimate Pendulum

- ✧ **The Finest Pendulum Ever Made.**
- ✧ **Precision-Weighted Brass Instrument.**
- ✧ **Wind-Tunnel Designed for Least Air Drag.**
- ✧ **Suspender Specially Designed with 157 Brass Ball-Joint Swivels for Maximum Freedom of Movement.**
- ✧ **Bonus... Black Velvet Carrying Pouch**

To receive your Ultimate Pendulum, simply fill in the coupon on reverse page, cut out and enclose with <u>your original store receipt printed with the name of the store</u> where the book was purchased. Your Ultimate Pendulum in a Black Velvet Pouch will be shipped to you by return mail. There is NO charge for this pendule.

Additional Ultimate Pendulums for gifts
may be purchased at $9.95 each
plus $4.00 for priority s\h for full order.

Note: To receive your Free Ultimate Pendulum <u>you must send original store receipt</u> and this coupon from the reverse page of this book.

SORRY NO PHOTO COPIES
will be honored. Thank You!

Coupon for the FREE Ultimate Pendulum

❑ Please send my FREE Ultimate Pendulum. I enclose my original store receipt.

❑ Yes! I would like to have ___ additional Ultimate Pendulums. *Limit 3 per order.*
I enclose $9.95 per additional Pendule plus $4.00 shipping. <u>Please make check to Powell Productions.</u>

❑ Please charge to my credit card # _____
_____ Expire Date _____

❑ Please send me your FREE catalog.

❑ I have never used a pendule before,
   this will be my first time.

❑ I have used a pendule before with great results.

❑ I have used a pendule before with
   less than great results.

❑ I would like to attend a *Taming The Wild Pendulum Seminar.* Please send dates and location.

Name_____

Address_____

City_____ State_____

Zip Code _____

Phone (_____) _____

FAX (_____) _____

**Mail to: Powell Productions
P.O. Box 2244, Pinellas Park, FL 34664-2244**

TAMING THE WILD PENDULUM

By Dr. Tag Powell and
Dr. Judith Powell

TOP OF THE MOUNTAIN PUBLISHING
Pinellas Park, Florida 34664-2244 U.S.A.

All rights reserved. No part of this book may be reproduced, utilized or transmitted in any form or by any means, electronic or mechanical—including photocopying, recording on any information storage and retrieval system—without written permission from the publisher, except for brief quotations or inclusion in a review.

TOP OF THE MOUNTAIN PUBLISHING
P.O. Box 2244
Pinellas Park, Florida 34664-2244 U.S.A.
SAN 287-590X
Fax (813) 391-4598
Phone (813) 391-3843

Copyright © 1995 by Dr. Tag Powell and
Dr. Judith Powell

Library of Congress Cataloging-in-Publication Data
Taming The Wild Pendulum/Tag Powell and Judith Powell.
p. cm.
Includes bibliographical references and index.
ISBN 1-56087-057-5 : $19.95
1. Fortune-telling by pendulum. I. Title
BF1779.P45P68 1995
133.3'23—dc20   94-11030 CIP

Cover and text design by
Powell Productions/Marcos A. Oliveira

Manufactured in the United States of America

Table Of Contents

LIST OF ILLUSTRATIONS 5
ACKNOWLEDGMENTS 7

PART I

INTRO: CAPTURING THE WILD PENDULUM 11
1. CREATING YOUR WILD PENDULUM 15
    Your Perfect Pendulum 18
    How To Choose That Special Object For Your Pendulum 21
2. TRAINING YOUR WILD PENDULUM 25
    Yes! 28
    No! 29
    Double-Checking Your Pendulum 30
    Practice... Practice... Practice... 31
    How To Determine If Conditions Are Right? 32
    To Be Or Not To Be... What Is The Question? 33
    Beware The Question Trap 35
3. HOW AND WHY DOES YOUR PENDULUM SWING? 41
    Is the Pendule Controlled by "Spirits"? 42
    Who Puts The Swing In The Pendule? 43
    What Controls The Accuracy Of The Pendule? 44
    Is The Pendulum Always Right? 45
    Judi's Solution—Don't Peek! 45
    Living With Your Pendule 47
4. THE MIRACLE OF TELERADIESTHESIA
   (MAP DOWSING) 49
    Map Dowsing To Locate A Missing Person 50
    Sometimes You Do And Sometimes You Don't 51
    4 Steps To Map Dowsing 53
    Discovering Deadly Noxious Rays 56
    Eliminating "Noxious Rays" 58
    How To Discover A Real Buried Treasure 61
    Your Success Log 62
    Should You Charge For Your Pendulum Work? 63
    Tithing To The Person Who Helped You 64

PART II

5. SOLVING WITH THE BASIC POWELL MULTIMETER 67
- Just What Is "It"? 68
- Exploring The Powell MultiMeter 71
- Using The MultiMeter Test Mode 72
- How To Number Crunch With A One-To-Ten" Reading! 73
- How To Set Your MultiMeter For Percentages % 77

6. DETECTING ILLNESSES: SWINGING INTO HEALTH! 81
- The Legality Of It All 83
- Testing For Better Health: Too Much Or Too Little 84
- Using The Health Band 85
- Testing, And More Testing 88
- Penduling With Body Illustrations 90
- Heath-Check Questions 90

7. PENDULING FOR ALLERGIES 103
- Three Testing Techniques For Foods And Liquids 104
- Questions, Questions, And More Questions 108
- Allergies Are Not Just To Food! 109

8. BOBBING WITH YOUR PENDULUM-ON-A-STICK! .. 111
- How Do You Use It? 112
- Training Your "Happy Puppy" 114
- Creating Your Bobber-Dobber 115
- Experimenting With Your Bobber 116

9. BOBBING FOR LOTTO WITH JUDI 121
- How To "Bob" For Winning 123
- Lotto Numbers 123
- Using Your Pendule For Lotto 126
- Other Places to Increase Your Luck 129
- To Gamble Or Not Gamble 129

10. DISCOVER YOUR PERFECT MATE, FRIEND, PET OR BUSINESS PARTNER! 131
- Quick-Check For General "Compat" 136
- Percentage Of "Compat" 137
- Nine Buttons For A Perfect Human Fit 138
- Personal 139

  Physical 140
  Mental 141
  Emotional 141
  Social 142
  Family 142
  Spiritual\Beliefs 143
  Work\Career 144
  Financial 145
  Gathering More Data 145
  It's All in the Percentages 147
  Fill In The Blanks 148
  Use Your Common Sense 148

11. TRAVELING IN YOUR TIME MACHINE! 151
  Looking Into The Future 153
  Predicting The Elections 154
  Discovering the Past 158
  Your Present Past 158
  Discovering The Beginning Of A Problem 158
  Discovering Your Past Lives 161
  What Is The Value In Past Life Regression? 161

12. THE MAGIC DOORWAY: OPENING YOUR CHAKRAS 163
  The Interconnection Of Physical, Mental, And Spiritual 167
  Testing Your Energy Levels In 3 Steps 168
  1. Root Chakra: 170
  2. Spleen Chakra: 171
  3. Navel (Solar Plexus) Chakra: 171
  4. Heart (Thymus Gland) Chakra: 172
  5. Throat (Thyroid) Chakra: 173
  6. Brow (Pineal) Chakra: 173
  7. Crown (Pituitary) Chakra: 174

WHERE DO YOU SWING FROM HERE? 177

RECOMMENDED READING 179

RESOURCES 181

ABOUT THE AUTHORS 182

BIBLIOGRAPHY 186

INDEX 188

List of Illustrations

DIAGRAM 1. Three major dowsing tools 17

DIAGRAM 2. Items which can make handy pendulum 20

DIAGRAM 3. Various ways to hold your pendulum ... 27

DIAGRAM 4. Map of St. Petersburg, Florida 60

DIAGRAM 5. Basic Powell MultiMeter 70

DIAGRAM 6. Powell MultiMeter with Health Band ... 86

DIAGRAM 7. Female body. 92

DIAGRAM 8. Male body. 93

DIAGRAM 9. Muscle level of human body. 94

DIAGRAM 10. Skeletal level of human body 95

DIAGRAM 11. Interior head and neck 96

DIAGRAM 12. Throat, respiratory, digestive systems . 97

DIAGRAM 13. Heart and main arteries, cross section . 98

DIAGRAM 14. Digestive system, close-up view 99

DIAGRAM 15. How to hold the bobber 113

DIAGRAM 16. Cross-section of bobber 118

DIAGRAM 17. Powell Compatibility Palette 135

DIAGRAM 18. Powell Time Machine 152

DIAGRAM 19. Powell Minute MultiMeter 157

DIAGRAM 20. Chakras Chart 165

Acknowledgments

We wish to show our grateful appreciation to the American Society of Dowsers for keeping "the word" out there for the public.

Thank you to our talented editor, Yvonne Fawcett, for finding just the "right" words.

And a warm thanks to all the students attending our seminars around the world. For their eagerness and enthusiasm in keeping the pendulum alive!

A FEW OF THE OTHER BOOKS & TAPES

BY DR. TAG POWELL

Silva Mind Mastery For The '90s (with Dr. Judith Powell)
Think Wealth: Put Your Money Where Your Mind Is!
As You Thinketh (with James Allen)
From Poverty to Power (with James Allen)
ESP For Kids: How To Develop Your Child's Psychic Abilities (with Carol Howell Mills)
Slash Your Mortgage In Half
Inner Power With NLP (with Dr. Judith Powell)
Mega Power Learning Audiocassette Kit
Think Money Audiocassette Kit
The Secrets of Holographic Visualization 2-Tape Kit
Silva Master Mind Seminar... 8-Audiocassette Series
Super Subliminals Plus... 8-Audiocassette Series

BY DR. JUDITH POWELL

Silva Mind Mastery For The '90s (with Dr. Tag Powell)
The Science of Getting Rich (with Wallace D. Wattles)
The Science of Becoming Excellent (with Wattles)
The Science of Well-Being (with Wallace D. Wattles)
A Date With Destiny
Discover Your Perfect Soul Mate Tape Kit
The Magic of Color 2-Tape Action Kit
Color Healing Audiocassette
Love Yourself 2-Tape Action Kit
Balanced Living 8-Audiocassette Series
Up Your Energy Audiocassette

PART I

TAMING THE WILD PENDULUM

Taming The Wild Pendulum

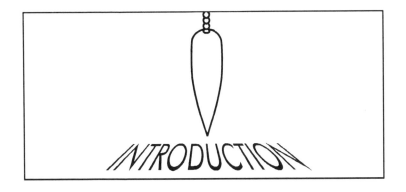

Capturing
The Wild Pendulum

If this book is your first step into the world of *using your inner-conscious mind*, some of these ideas may seem "too far out" from your current way of thinking. This information may even stretch your envelope of believability. But if you truly desire to grow and succeed in life, it may be beneficial for you to look upon these "different" ideas as a metaphor for a higher functioning of learning.

Too many accept only those thoughts within their current thinking sphere and reject the rest. Rather than reject any idea in this book that seems a bit "wild," put the idea aside for later review after you have achieved some success with the basic pendulum techniques.

Within these pages, you will learn how to choose or create and then use your pendulum or pendule (these words will be used interchangeably throughout this text). By using the pendule as an outer-conscious tool, you will discover how to use your inner-conscious mind to aid you in decision-making in your personal life, your business life.

It has been shown by research that the most successful business people score high in intuitive testing. We personally know many eminent, successful individuals who use this instrument in their decision-making process.

First, they use the analytical processing mind to weigh the facts. When there are not enough facts known to clear a path to a final decision, they take a second step: *intuitive* processing through the use of the pendule, gaining insight to knowledge unknown to their outer senses. By using both the right and left hemispheres of the brain—considered the inner- and outer-conscious minds, respectively—these individuals are able to release increased creativity and greater problem-solving abilities.

The pendule can be a tremendous help if you are one of those tormented by indecision, or have any problems at all in being able to pick and choose—whether it's a house, a tie, a mutual fund, an entree... or a mate!

You will learn to use your pendule to find lost objects, make investments, and maybe pick winning lottery numbers! You will learn to "dowse" a map in order to find oil, lost treasures, even missing persons. You will explore methods to improve your health and the health of your family and friends—to *psychically* detect illnesses, to check for allergies, for nutritional and vitamin deficiencies.

We have also learned of psychotherapists and psychiatrists who use the pendulum as a *"facilitator."* It helps a patient break through after a traumatizing event when life has left them so blocked the memory is buried deep in the inner conscious. The therapist questions; the pendule, in the hand of the listening patient, answers.

One word of advice: It's better to read this book from the beginning, not jump in where something tantalizes you, because it is a learning, building process. It explains *why*, it shows you *how* as you go along. Should you leap ahead, you could confuse not only yourself but the pendule! It needs to build too. So tame the Wild Pendulum.

Good hunting!
Drs. Tag and Judith Powell

Taming The Wild Pendulum

Creating Your Wild Pendulum

"A body hung as to swing freely to and fro..." is how *Webster's New World Dictionary* describes a pendulum. *(L. pendulus, pendere* to hang).

The study of this swinging object has been a favorite throughout history, and today it's even the subject of scientific research. The pendule has been used in areas as varied as being the control mechanism for pendulum clocks, to recording the rotation of the earth—as seen in the *Foucault pendulum* in the National Museum of American History in Washington, D.C.

The pendule, along with the rod and the wand, is one of the oldest metaphysical tools, a psychic instrument proven effective for thousands of years. As eyeglasses are an instrument to improve eyesight, the pendule is a tool to increase one's seeing with the *inner vision* of the psyche or mind.

The pendule in plumbob-shape is used even today not only in surveying but in dowsing the land. Along with the traditional forked stick or Y-rod, and the bent metal or L-rods, the *plumbob* has earned its place in the dowsing field. (see Diagram 1)

What really is *dowsing?* Merely the use of some physical tool to exercise the inner faculty of *intuition* or *right hemisphere* of the brain, to obtain an answer unknown to the outer *logical* senses or *left brain hemisphere.* Dowsing is using a tool to define one's intuitive feeling. The pendule has the advantage of being the most portable of all dowsing instruments, awaiting your command in the corner of your pocket!

The most versatile of psychic instruments, the pendulum will perform the functions of almost all the other psychic apparatus combined.

The use and beliefs about this mystic tool are as varied as there are people. Nor are the methods you are about to learn the only ways in which to use this instrument. If you are currently using a pendule with good results, stay with that method. But if you are not

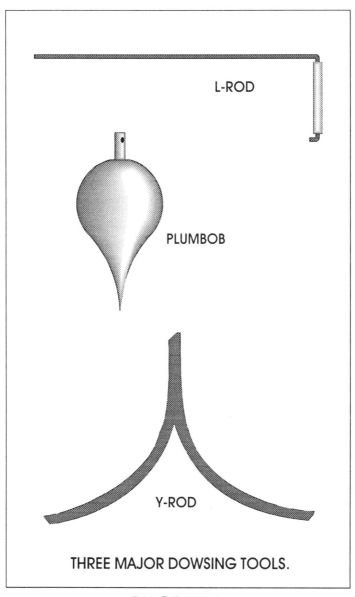

DIAGRAM 1

getting the results you want; or if you never used a pendule before; or if you want to expand beyond... read the *Wild Pendulum* and grow. Reap the gains from this new knowledge!

Your Perfect Pendulum

When you *purchased* this book you should have received *The Ultimate Pendulum* (or a coupon in the back of this book to send for this precision tool, free of charge).

To give you "a near-perfect performance," I (Tag) designed a brass teardrop pendulum in classic shape. It was specially constructed for suspending a brass bead chain with ninety-four swivel points for freedom of movement.

Although a great deal of research went into my design of this Ultimate Pendulum, I believe a homemade pendule will work quite well. So if you are waiting for your Ultimate Pendulum to arrive, and you want to get started right now, you have two choices: 1) buy a pendule at your local metaphysical or New Age bookstore; or 2) much more fun, make your own!

If you are buying one many New Age bookstores carry interesting selections, costs ranging from $3.00 to $150.00. Until developing the Ultimate Pendulum to accompany this book, my favorite was a "Brass Bullet Pendulum" and is available from *The American*

Creating Your Wild Pendulum

Society of Dowsers, Inc. (A.S.D.), address in Resources section.

If you're creating your own—the most popular shapes are the *tear drop* and the *streamline* (what a surveyor uses). But a good working "homemade" pendulum can be made from almost any small object attached to a string: choose a bead, a button, a charm, a pendant. Though usually round, conical, or pear-shaped, an interesting pendule can be made from some of today's mechanical parts, such as a metal washer—or try a key or a nail! The material itself may be anything from plastic to gold, from lead to crystal. Even a tea bag can make a good pendulum. (see Diagram 2)

The *suspension system* can also be made from anything—thread, string, dental floss or best, a chain. Chains have the advantage of being free from "drag" that often distorts a reading with a string-held pendule. As for length, a good rule of thumb is *"the lighter the pendule body, the shorter the suspender length,"* usually from three to nine inches.

So have some fun looking through an old jewelry box to find a "special object," to be the *bob* in your plumbob—an old pendant, earring, a ring. If you choose to use a ring that you usually wear, at least it will always be handy for emergency dowsing to obtain unknown answers!

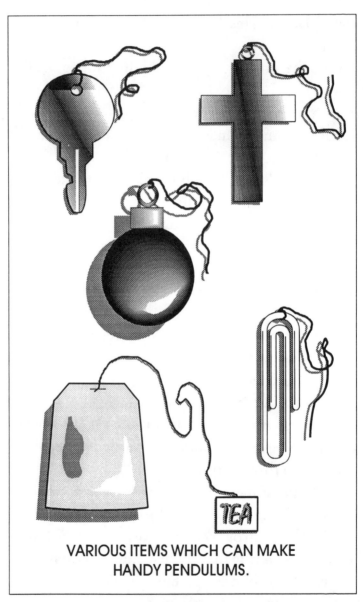

VARIOUS ITEMS WHICH CAN MAKE HANDY PENDULUMS.

DIAGRAM 2

How To Choose That Special Object For Your Pendulum

1. Select several items which you believe will make a good instrument. Place the objects on a table in front of you and sit in a comfortable chair within easy reach of them. Take a minute to close your eyes and relax. Take three deep breaths and exhale slowly as you tune into your *inner self*.

2. Next, rub your hands together to increase their sensitivity. (The friction removes dead cells from the palms and increases circulation.) Choosing one of the objects, pick it up and lightly clasp your hands around it. *Feel* or *sense* its *vibrations*.

You may interpret the vibration as a tingling feeling; a cool or warm sensation, a "different" feeling; or a special *knowing*, which will distinguish one article from another. To increase your sensitivity, close your eyelids and take a deep breath and exhale slowly and tune into your Higher Self. Keep testing and shifting from one object to another until you decide which is to be your "special pendulum."

3. Now decide on what type of suspension material you will use. Again, it's your personal preference—a thread, a string, a chain, even in emergencies, a scarf!

If you choose a chain, be sure to use a fine chain with *round loops*. A large chain, especially one with odd-shaped loops, can constrict or alter the pendulum's directional movement. The length of the suspender can vary. The longer lengths working best, of course, with heavier pendules. Our favorite length is *six inches*—approximately one inch held between the fingers, giving a five-inch swing.

4. Find a small pouch, purse, or pillbox to carry your special pendule around with you (a black velvet pouch is provided for you with our Ultimate Pendulum). Some feel the pendule should be carried on your person for at least two weeks before using in order to absorb your *personal energy frequency*, to attune it to your psyche.

We question the two-week waiting period. Your pendule is a tool like a hammer, and a carpenter would not wait two weeks before using a new hammer. Begin to use your pendulum right away.

Creating Your Wild Pendulum

You will want to carry your pendulum with you at all times as you will find a great many ways to use it: questions to answer, decisions to make, new doors to new worlds to open. The more you use your pendule, the more you will believe in your intuitive mind. And this is so.

Taming The Wild Pendulum

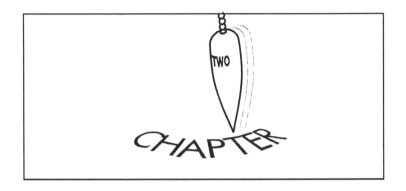

Training Your Wild Pendulum

Some dowsers insist you allow the pendulum to *program* itself—letting it decide how to swing and answer. But how would you know what the pendulum is telling you? We believe the best answers are determined by *you* programming your own pendule.

The method you are about to learn has been successful for thousands of people we have trained. It has also gained better results for individuals who in the past searched for answers but received confusing or inaccurate readings. The final answer is, of course, use what works for you!

Also, let it be noted that programming your pendulum in this specific manner will help create a more universal standard for pendule operation. For example, most screws today are made with a right-hand thread. Because of this standardization, anyone can tighten a screw by turning it to the right and loosen a screw by turning it to the left. Imagine the confusion if each screw manufacturer made screws right- or left-thread at random!

By standardizing pendule operations we not only make training easier, we open the door to much more complex procedures (as compared to only a Yes or No answer)—which you will learn in later chapters using the *Powell Pendulum MultiMeters*©.

Let's now begin to "train" your pendulum.

STEP ONE: Hold the suspender of your instrument between your thumb and your first two fingers. (see Diagram 3) My favorite position is "D." We prefer about five inches from the top of the suspension to the pendule body, but try different lengths until you get what feels best for you. Now brace your elbow on a table or in the cup of your other hand, as at times you will be working for extended periods.

STEP TWO: The next step is to program or *imprint* your pendulum. This is

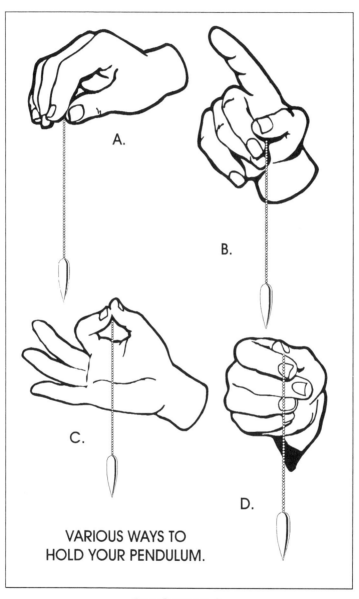

DIAGRAM 3

important—to establish a working "language" for successful communication between your inner conscious and your pendulum procedure. Remember, your pendule is just the tool, the telephone, for your inner conscious to talk to you where you usually live, in your outer-conscious world.

Yes!

First, program your pendulum to understand what "Yes" means. Deliberately swing the pendulum out away from your body and then back toward you—all the while thinking "Yes." Keep the pendule swinging back and forth while thinking Yes! Yes! This back-and-forth movement is a close equivalent to nodding your head in a Yes fashion. So this first time, nod your head "Yes" while swinging the pendulum... and at the same time thinking or saying aloud, "Yes."

Sense it! Feel it! Pretend it! Know it! Keep swinging the pendulum back and forth thinking "Yes! UH-UH!" Think positively! Think absolutely! Think "Ya! Si! Da! YES! YES! YES!" Relax and think Yes.

Close your eyes, or blur your vision, or enter a relaxed meditative state. Take a deep breath and exhale slowly. Keep the pendulum swinging back and forth, to and from the body, while thinking YES! YES!

YES! While thinking Yes, continue nodding your head Yes, up and down... all to help reinforce the Yes answer. Repeat this process for a full minute.

You have now programmed your pendulum (and your *inner conscious*) as to what constitutes a YES answer.

No!

Now is the time to become noble. (To be noble is to be able to say "NO" without guilt.) Encode your psychic tool to understand what constitutes a "No" answer. Purposefully swing your pendulum from left to right, side to side, while thinking "No!" Keep the pendulum swinging to each side, while thinking NO—shaking your head from side to side—saying aloud, "NO! NO! NO!"

Sense it! Feel it! Pretend it! Know it! Keep swinging the pendulum from side to side, thinking NO! Think negatively! Think "Nein! Nyet! Nada! No! No! No!" Relax and think "No."

Again close your eyes, or blur your vision, or enter a relaxed meditative state. Take a deep breath and exhale slowly. Keep the pendulum swinging from side to side, right to left. While shaking your head from side to side... think NO! NO! NO! Repeat this process for a full minute.

You have now programmed your pendulum (and your *inner conscious*) as to what constitutes a NO answer.

Double-Checking Your Pendulum

Check your pendule by phrasing a question which has to have a Yes answer, such as: "Am I at home?" (wait for answer) "Am I eager to learn about penduling?" (wait for answer) Now phrase a question which has a No answer: "Am I in Egypt?" (wait for answer) "Am I standing on my head?" (wait for answer)

Once you ask these questions, and get the correct swing movement or response, you will have completed the basic encoding for Yes and No answers.

A word of caution: You have programmed the pendulum to swing in a specific manner for the Yes and No answers—*do not switch these patterns to denote something else.* This will only confuse the pendulum, your inner conscious, and you. Your results will be confusing, or invalid. Always keep the same program—always keep the same movements for Yes and No.

For you "old swingers" who have swung the pendule before—with YES being from right to left where NO should be (because that is the universal head shake for YES)—you will have to work and concentrate a little harder to eradicate the different program. Having done it the reverse, your answers may have been the reverse! So try it—imprinting your Wild Pendulum the natural, universal way—and we promise you more accurate, less confounding answers.

Practice... Practice... Practice...

Your pendulum is now ready to go to work. Start thinking of all the ways you can use your new psychic instrument. As examples: dowsers report that the pendulum is the best tool to test if you are allergic to certain foods; to test the comparative health of your body organs; to test if you really need vitamins and if so, how many of which ones. This Health Testing you will learn to do with the Basic Powell MultiMeter©, in Part II. For right now, let's stick to the simple questions which require only a Yes or No answer.

Carry your pendule with you, and use it every chance you can. You will sometimes get the correct answer, other times you won't; and still other times you won't know what answer you're getting! Remember, YOU are probably still controlling the swing, rather than your inner conscious doing it.

How do you go about turning over the "control" to your all-knowing inner mind? Build up your confidence. Practice for at least three consecutive days just getting Yes and No answers. Practice so that you learn to relax—not questioning your results—or worse, trying to predetermine them!

Practice... practice... practice... and you will soon be ready to learn how to swing for lotto numbers; how to map dowse; how to use the three Powell Multi-Meters©.

How To Determine If Conditions Are Right?

As in all things, sometimes the conditions or the timing is not right. This may be checked by simply asking your pendule, "Are the conditions ripe for testing?" If the answer is No, then ask, "Should I try to correct the conditions? If the answer is again No, you should put away the pendule until conditions are "ripe."

But if the answer is Yes, you have the chance to play twenty questions to find out just how to correct the conditions.

Remember: the more specific the question, the better the answer.

First, check out your surroundings:

"Is the present environment distracting?"
(wait for answer)
"Should I leave this area?"
(wait for answer)

"Am I being distracted by someone in this room?"

(wait for answer)

"Is someone, or something, in this room so NEGATIVE that it can influence me or the pendule?"

(wait for answer)

This concept is so important that it should be considered in all your life dealings, business and personal. Strong negativity coming from your associates, your family or your environment can pull you and your psyche down... down so far that some actually become ill.

As a new *pendulist* you may feel strange using a pendule in the presence of others or in a public place. One solution may be a quick retreat to the restroom for solitude and peace. But don't make those trips too often as others may think you have a kidney problem!

To Be Or Not To Be...
What Is The Question?

Let's try a few basic questions to start exercising your pendule *and* your inner conscious:

- ✧ "Should I wear the blue tie/dress to the meeting/date/party?"
- ✧ "Will I be happy in this job?"

- ✧ "Will I have job security here?"
- ✧ "Am I on track to reach my goal?"
- ✧ "Am I in the right "space" for my physical/mental well-being?"
- ✧ "Is this the person I want to spend the rest of my life/vacation/evening with?"
- ✧ "Is this person safe for my physical/psychical well-being?"
- ✧ "Is my relationship in trouble?"
- ✧ "Is my lover/spouse cheating on me?"

These last questions are getting serious in today's world. A pendulum is handier, cheaper than a private detective!

- ✧ "Should I change my diet?"
- ✧ "Am I getting the proper exercise for vibrant health?"
- ✧ "Am I getting enough sleep?"
- ✧ "Am I thinking the right thoughts to reach or maintain vibrant health, in mind and in body?"
- ✧ "Should I think more positively?" (You may assume you do, but too many of us get into such a negative-thinking-habit rut that this definitely needs to be checked out—and often!)
- ✧ "Should I stay away from, whenever possible, negative places and negative people?"

So these questions seem overly obvious to you? You'd be amazed at how many 'smart people' ignore these very basic prescriptions for a happy, healthy, sane life.

When you ask these questions, should you say them aloud? *It is usually more effective if the question is asked out loud.* By framing it, you *focus* on the precise question, *clarifying* your thoughts. Those times when it is prudent for observers not to hear, just say it to yourself mentally, but do mentally frame a *complete* question. Incomplete or vague questions get incomplete or vague answers. Then you blame your pendulum and lose trust in it and in yourself.

Beware The Question Trap

The hardest part of learning to use the pendulum isn't *getting the correct answer,* but *thinking the correct question!* For any possible scenario, ask always in a clean, precise line of thinking/questioning. It is vitally important to *ask only one question at a time:*

Incorrect: "Should I move to another house or stay where I am?"

You cannot get a single Yes or No answer, because you are asking two questions at once.

Correct: "Should I move from this house?" If the answer is Yes, ask: "Should I move this year?"

If the answer is No, ask: "Should I stay where I am?"

Always give the pendule *time* to answer. If it just swings inconclusively in a circle, you in your inner mind are still too uncertain, confused. Wait a while, then try again.

Let's play out more of the above scenario about moving. Pretend the answer to the moving question was Yes. You would continue to ask further questions, such as:

"Should I sell this house?" (No)
"Should I keep this house as a rental property?" (No)
"Should I keep this house as a second home?" (Yes)
"Should I buy another house?" No
"Should I rent another house?" No
"Should I buy a condominium?" No
"Should I buy a town house?" Yes
"When is the best time to make this buy: within three months?" No
"Between three and six months?" Yes

You would then continue the questioning of the pendulum to pinpoint the exact month of the move (use the Time Travel MultiMeter©) and the location of where to purchase your new house (engage in

Map Dowsing) also at what price! (For this last important consideration, you may check your Basic MultiMeter©!)

Look for loopholes in your questions. A careless or ill thought-out question can produce a false or misleading answer. So form your questions in a simple yet complete manner. Long ramblings—trying to cover all bets in one question—can cause loopholes, indecision, wrong directions. In most cases, *several questions will be needed to discover a final and correct answer.*

Sometimes we do not ask the right questions; other times we do not ask *enough questions* to cover all the possibilities. One example I'll never forget concerns a woman we trained to use this psychic tool.

She had lost one earring. The earrings were diamonds, an anniversary present her husband had scrimped and saved for a year to be able to purchase. After a thorough physical search produced no results, the couple decided to do a psychic search—to use their inner powers, and test with the pendulum.

"Is the earring in the house area?" they asked.
Answer, NO!
"Is the earring in the yard area?"
Answer, YES!

Since their property was large, they wanted a more specific location. They asked:

"Is the earring in the front yard?
YES!
"Is the earring within 100 feet from the house?"
YES!
"Is the earring within 50 feet from the house?"
YES!
"Is the earring within 25 feet from the house?"
YES!
"Is the earring within 15 feet from the house?"
NO!
"Is the earring within 20 feet from the house?"
YES!
"Is the earring on the right side of the front yard, when facing the house?"
NO!
"Is the earring on the left side of the front yard, when facing the house?"
YES!
"Is the earring in the driveway area?"
YES!

Great. They had determined the earring was between 15 and 20 feet from the house and in the driveway—a driveway that consisted of loose crushed stone! They searched very carefully, even moving their pickup truck to check beneath it. No diamond earring.

Okay, back to the pendule to try to pin down the location in that loose crushed stone.

"Is the earring within 16 feet from the house?"
NO!
"Is the earring within 17 feet from the house?"
NO!
"Is the earring within 18 feet from the house?"
NO!
"Is the earring within 19 feet from the house?"
NO!
"Is the earring within 20 feet from the house?"
NO!

Now they had a problem. First the pendulum said the earring was within 20 feet of the house, now it was saying NO, it's not within 20 feet. They started over:

"Is the earring within 30 feet from the house?
YES!
"Is the earring within 25 feet from the house?"
NO!
"Is the earring within 26 feet from the house?"
YES!

The couple went back to the driveway and carefully measured 25 feet from the house. To search through the stones, they had to move the truck again.

While the husband was searching the wife had an idea. Back to the pendule she went and asked:

"Is the earring in the truck?"
YES!!!

They had, of course, already looked on the floorboard, under the seat and in the back of the truck—with no results. The wife asked:

"Is the earring under the seat?"
NO!
"Is the earring in the seat?"
YES!

Upon removing the seat, they found the diamond earring. It had fallen off her ear and slipped down into the seat separation where it had remained hidden.

Conclusion: Sometimes the most obvious question is missed!

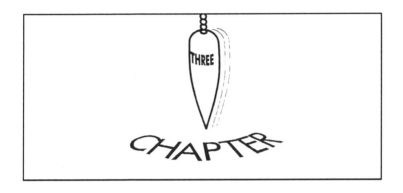

How And Why Does Your Pendulum Swing?

My (Tag's) first "Magic Pendulum" was purchased by mail when I was about ten years old, from the Johnson Smith Company. It was a mail order supplier of tricks and jokes, wonderful and unusual items. (They are still in business today.) The instructions that came with the pendulum stated you could tell the sex of a chicken egg by holding the pendulum above the egg. As I could never figure out why anyone would really care about the sex of an egg, I quickly lost interest in the "Magic Pendulum."

The next time a pendulum came into my life was when I was fourteen and selling the *Grit Newspaper* door to door. One of my customers tried to interest me in a complicated allergy tester. This was truly an intriguing setup: a beautiful leather briefcase with tiny compartments for about a hundred little vials of minerals and foods that people might be allergic.

These vials were to be placed on the inner fold of the arm at the elbow, while the other hand held a pendulum over a target-drawing. At the time, I felt this allergy detector was probably as useful as determining the sex of a chicken egg. Looking back years later, I realized it was a beautifully made professional instrument, and I have often thought of making one. But I know full well that the way we use the pendulum to detect allergies (Chapter 7) is a handier and simpler way. It's just that I love beautiful gadgets!

Is The Pendule Controlled By "Spirits"?

Do not confuse the ability to utilize the inner conscious with being supernatural. Yes, pendule usage is considered by some as metaphysical. However, today, these so-called metaphysical phenomena are not only being explained by science but are often being reproduced under laboratory conditions! Quantum physics has opened the door to a new understanding of the world around us, and above us. The pendulum is just a tool, like a key, that opens another door.

Using the pendule to make decisions is no more supernatural than using your memory to call up a word for a crossword puzzle, or using your ingenuity to create a new product or a poem or a sonata.

Who Puts The Swing In The Pendule?

Critics of the pendulum have long said that *the control of the movement* was caused by the pendulist. Today, I realize that is partially correct. The point the critics are missing is that, although the pendulum movement IS caused by the very slightest movement of the muscles—it is NOT under the "conscious" control of the pendulist, but rather controlled in some manner by the transference of thought-energy *through* the physical body.

There is another kind of pendule movement which is not touched by the human body in any manner. Here the movement is caused by *psychokinesis.* Two types of Psychokinetic Independent-Suspension Pendulums are detailed in my friend G. Harry Stine's book, *Mind Machines You Can Build*. His book contains fascinating devises which you can actually build and affect with just your mind. These "instruments" function entirely differently from the ones described here in *Taming The Wild Pendulum*.

In your hand-held pendulum, the power is also NOT in the instrument but in the power of your mind—your inner conscious that "intuitively" knows —and then transfers data to your outer conscious through minute muscular movements.

These muscular responses are coming from the *thought-energy* of that inner mind. An energy that travels through the nervous system as impulses... through the ulnar nerve to the fingertips... causing the almost imperceptible action that makes the pendule move beyond conscious, beyond physical control.

Of course, we can only theorize about how it exactly works, while keeping in mind that it *does* work. The latest research on the functions of *the right* (inner conscious) and *the left* (outer conscious) *hemispheres of the brain* opens the door to a better understanding of this phenomenon.

What Controls The Accuracy Of The Pendule?

The correctness of your pendule is as accurate as you will allow it to be. That is, you need to allow the pendulum to swing on its own when asked a question. The problem with most beginning pendulists is that the left brain hemisphere (outer-conscious level) wants to remain in control.

This left "analytical" brain has been making the decisions through most of your adult life. Due to years of "indoctrination," this part of your brain will want to either take control of the pendule or try to prove it useless by forcing wrong answers. The secret is to RELAX while using this mind-tool in order to block out the negating analytical mind. Again, practice... practice... practice.

Is The Pendulum Always Right?

No, of course not. There is probably nothing or no one who is always right. On the positive side, the odds are good, and getting better the more you swing! In the beginning, if you are like most people, you will be using some conscious control over the swing, and will force (either consciously or unconsciously) the answer you want (or don't want, if you are trying to prove the pendule wrong). The solution may be to not tackle any serious problems in the beginning, until you feel you have confidence in the pendulum's answers.

Judi's Solution—Don't Peek!

I (Judi) have discovered the secret to obtaining more accurate answers from the pendulum. A way to get rid of that doubting, distracting, misleading outer conscious... that overly analyzing mind that many beginners are caught in, and even old-timers! I have shared this secret with longtime pendulists, and they have found my "trick" to be extremely helpful.

It particularly helps people who are concerned that it is *they* who are moving the pendulum to say YES or NO simply because they want that answer. So to get your *limited* outer-conscious ideas out of the way, and to tap into your *unlimited* inner-conscious wisdom for the best answer, try the following:

1. Get two identical sizes of paper (heavier paper is better so you can't peek through. Identical business cards are perfect for this).

2. Write "YES" on one card, "NO" on the other card or paper.

3. Turn the two backwards, and mix them so your outer senses forget which is which.

4. Place their yes/no side down on the table, about four inches apart.

5. Phrase your question for a simple Yes or No response. ("Am I going to Paris?") *Ask your question aloud.*

6. Now hold the pendule over the paper on the LEFT and ask, "Is this one the correct answer to my question?" (wait for response) If the pendule gives a YES response for that card, push it forward a couple of inches. If the response is NO, leave it in place. Don't peek!

7. Now move your pendule over the paper on the RIGHT and ask, "Is this the correct answer to my question?" (wait for answer)

8. Okay, you can now turn over the paper that had the *pendule response* of Yes. Whatever that paper reads is your correct answer. If the pendule swung Yes over the paper that reads NO—you aren't going to Paris. Sorry.

9. But if the pendule just happened to swing Yes over both of the cards, giving you both a YES and NO answer, you should ask more questions, qualify it. Ask if you should rephrase your questions more specifically. Are you going next year? NO. Are you going sometime in your lifetime? NO. Are you going in your dreams... by reading travel books... by seeing it on TV? YES.

Living With Your Pendule

We both use the pendulum almost daily, in our business and in our personal lives. We trust the answers and generally follow its advice. There are those *rare* times, however, when we feel it necessary to disregard an answer. The pendule, we feel, like the head of the hammer, does not always squarely strike the nail!

Part II of this book goes into some fascinating, yet easy and quick, techniques you can do to get more out of your pendulum than just a Yes or No. You will learn how to get percentages and amounts and other involved solutions and answers—even how to pick the right mate, friend, pet, employee, employer!

For now, continue to build up your confidence getting Yes and No answers.

PRACTICE!

LIFE! It's better with a pendule!

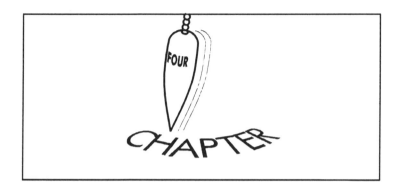

The Miracle Of Teleradiesthesia (Map Dowsing)

Most people have had the *Indiana Jones* dream. Finding a sunken ship, a buried treasure, a lost gold mine, or a prehistoric artifact. And often there is a need to find a lost person, a ship at sea, a downed airplane. *Cartographic, Map,* or *Distant Dowsing* with a pendulum has become one of the most effective methods to help the dreamers reach their goal; to help the frantic find their lost loved ones.

A dowser's preferred tools of the trade are an "L" rod (our favorite), the traditional "Y" forked stick, or one of the commercial devices. But these tools are used when the dowser steps foot on the actual property. What about distant places unreachable for reasons of economy, politics, war, or a host of other reasons and dangers. Many dowsers have neither the time nor the money to travel to a specific faraway spot. (Nor do they always know *what* specific faraway spot!)

If you are asked to dowse a place in a distant land or even the distant home of a friend, what can you do? The problem is easily solved. You, the pendulist, need to get a map or a drawing—and perform Distant Dowsing, or *Teleradiesthesia* (*tele* = distant; *radiesthesia* = sensing) as it was named by Master Dowser Vern L. Cameron.

If you are seeking sunken treasure, it is certainly easier, and cheaper to use a pendule to locate the general latitude and longitude before actually going aboard a ship with your "practical" dowsing tools.

Map Dowsing To Locate A Missing Person

Some years ago, I (Tag) was asked to find the location of a kidnaped businessman being held for ransom in Colombia, South America. Given only a small map to work from, I still was able to pinpoint a specific location. This information was communicated to authorities in Bogota.

The location I had noted on that small map worked out to an area about three miles square. It turned out that a search of that area was already in progress. However, before the search could be completed, the family paid the ransom, and the businessman was released, unharmed. The kidnapers were never found.

Although it came too late to help recover the kidnaped victim before paying the ransom, the police acknowledged that my dowsing expertise was correct—the man was being held within that three mile area. Had a larger map been available, I could possibly have been able to pinpoint the precise location, allowing the police to free the businessman. Perhaps save his money, and at the same time, catch the kidnapers.

> *Note: A small map can be used to define the general area. Then get a larger, more detailed map to pinpoint the specific location. Or, if no detailed map is available, at least enlarge on a photocopy machine the area you pinpointed as being "the spot."*

Many believe you can enhance your success if you have a photograph, or something personal, which belongs to the missing person. This object is often called a *"witness"* by dowsers.

*Sometimes You Do
And Sometimes You Don't*

Although most beginners have excellent results, sometimes they, and even the professionals, can be 100% off the mark. The answer as to why dowsing is occasionally "off" is a mystery. I know some experienced dowsers who believe that what they eat has a factor in their abilities. Certainly, if you are in poor health or are simply not feeling well (or not up to the mark!), it can affect your accuracy.

In past years, *The American Dowser Quarterly Digest* as held map dowsing experiments. A map is printed in the journal and members are asked to find a hidden object by using map dowsing techniques. Admittedly, they have had both positive and negative results. Some of the inconsistencies we believe can be attributed to the emotional connection to the search.

A person can usually be more effective in searching for a lost loved one than looking for a silver dollar hidden up in a tree! On the other hand, one can be hampered by too much emotional bias about the suspected location of a loved one.

As to finding treasure, that too can be hit or miss. The difference between need and greed often puts a fine point on it. Because of this, some of the most effective pendulists consider map dowsing a hobby

The Miracle Of Teleradiesthesia

or an adventure—they want to succeed, but are not thunderstruck if the location fails to provide the goal in mind. The reasons for success or failure, especially when looking for financial gain, can be very broad.

Again, success increases with practice, which increases your confidence, which increases your chances of more success. Coupled with a strong, true *desire* to succeed!

So what about a simple dowsing function like finding water? It is true, of course, that some professional dowsers do not always find water, but then many times certified geologists don't find water (or oil!) while using millions of dollars worth of *scientific* equipment.

4 Steps To Map Dowsing

The actual process for map dowsing or Distance Dowsing with a pendulum is a simple technique. It can be used to discover oil, gold, water, people. Even find the best place for you to work or live!

STEP 1: As in all of your pendulum experiments, the first step is to ask aloud, "Is this the right time to find the answer I am seeking?" If the answer is Yes, proceed to Step 2; if the answer is No, see page 32.

STEP 2: We assume you know your general destination enough to have chosen a map. Now reduce the search on that map to a specific area by chunking down to a manageable size. Divide the map into quarters or quadrants. You can do this by marking it with a pen or pencil. However, if you're like us, you may find the marks distracting, especially if you get a reading that falls on a line. We prefer to mentally divide the map, or to *fold the map into squares*.

To locate the area to be dowsed, you will need a pointer—pen, pencil, or large nail, will do fine. You could use your finger but a sharp point is better to zero in on a precise location.

STEP 3: With the pendulum in your right or dominant hand, the pointer in the other hand, *touch the pointer to the approximate center of the upper right quadrant*. Ask, "Does this map quadrant contain what I am seeking?" If you get a NO, move on to the next quadrant, moving always in a *clockwise motion*, touching with your pointer the center of each quadrant until you receive the Yes response.

When you get a YES, stop and divide this specific quadrant into a *subset of four quarters*, and repeat the process of touching the pointer to the center of each. This will help you find the location in the shortest amount of time.

You can continue to do this quartering until you get to a section approximately three or four inches square, a size easy to work with.

Continue to ask the pendule to give you a Yes when you are near the correct area. YES! You are there in the target area. Now "blow up" or enlarge that target area on a copier, so you can get greater details and accuracy. (Remember my Colombian map experience!)

STEP 4: On this enlarged map, let's narrow down even more the target area, working toward pinpointing the exact location. Starting at the *upper left-hand corner*, very slowly move your pointer across in a straight line *to the upper right-hand corner*. Drop down about a quarter- to half-an-inch (depending on the size of the map area) and with your pointer *now move in the reverse direction, from right to left* (you will be making a *"boustrophedon,"* or the old ox-plow pattern).

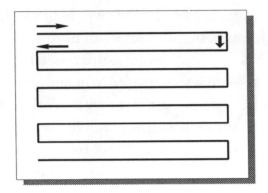

If you get a NO, continue past that to complete the row, then drop down, repeating the left-to-right, right-to-left process. When your pendule gives you a YES, mark it with a small yellow highlighter dot—that's your target area!

Discovering Deadly Noxious Rays

For several decades, pendulists have been dowsing houses and businesses and discovering what they called *noxious rays* or *noxious gases*. Many in the scientific community said, "Those quacks are wasting their time and ours. There's no such thing as strange rays and gases seeping up from the ground beneath the floors. They might as well be looking for bogeymen, ghosts, or something equally as silly."

It has only been within the last few years that science has "discovered" *radon gas* as the "bogeyman" destroying the health of thousands of people. These

The Miracle Of Teleradiesthesia

gases are apparently "seeping up through the floors of houses and office buildings" across the United States. Is radon gas the problem dowsers have been discovering for years? Is it just part of the "noxious gases" problem?

Now with distant dowsing techniques, you can check out the houses of loved ones miles away.

Just have them send you their floor plan. This can be an actual blueprint, or a hand-drawn layout of the rooms. A rough drawing will work; a scale drawing of course, with dimensions, will work even better.

Proceed as you did in Map Dowsing, but check each room on the plan very slowly, very carefully with your pendulum and pointer—because this concerns health, not just wealth! Be sure to use your pen or pencil as a pointer, and *use a highlighter to mark a dot* at any and all readings you get—you are looking for the *path* of the ray or gas.

Continue with the entire floor plan. In some rooms you may not get a reading. Just mark each spot where you do. Now connect the dots in order to view the path of the "noxious ray." If the path created by the dot pattern is not an obvious flow, it may be that the rays, rather than making a path, are just popping up

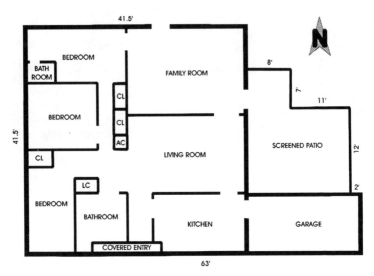

randomly in various places.

It could also mean you should go back and map dowse again the entire floor plan to check your readings. You might even find more locations. Connect the dots to see if they do make a path.

Eliminating "Noxious Rays"

Dowsers have ways to actually move or divert noxious rays or gases, and several books have been written on the subject (see Bibliography). There are many ways the two popular, simple-to-perform techniques *can split and divert a ray* to decrease its destructive power.

METHOD 1: One technique is to place a wire across the ray's path. An uninsulated lamp cord or even a coat hanger will work. This *diverting wire* can be taped into place and then covered with a rug (to keep everyone from tripping, or disturbing the *diverter*).

METHOD 2: Another technique is to take a three-foot length of heavy-duty aluminum foil. Fold the foil in half lengthwise, again and again, until it is a three-foot long, inch wide strip. Press it flat. Place it across the ray's path and tape in place. Cover with a rug.

Of course, if you are distant map dowsing for noxious rays, tell the person these diverting procedures. Have them choose a method, take action, then call you back when the diverters are in place.

Wait just a few minutes. Now pendule-check the floor plan again to see if the diverter has nullified the effects of the ray. If necessary, repeat the "diverting process" at another place along the ray's path. Try putting the wire or aluminum foil at different places in the house to discover where it can be most effective.

Does all of this sound like hocus-pocus? If it works, don't mock it! You'll know if it works by what your pendule tells you.

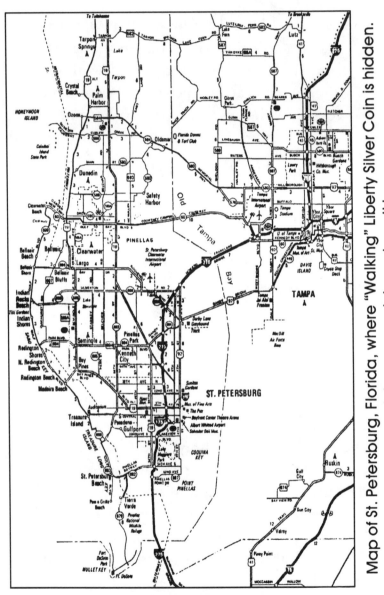

Map of St. Petersburg, Florida, where "Walking" Liberty Silver Coin is hidden. Ask your pendulum where it is.

DIAGRAM 4

Let it be added: noxious rays can be a mixture of various gases that *sometimes* includes radon. This "exorcising" technique used by dowsers works on the noxious gases and rays, but has *not* yet been proven to be effective on radon. If you have any reason to suspect radon gas in your area, it is always safer to send for a testing kit (see Resources section).

In the meantime, *dowse first*, then *divert*, and then *verify* by testing, if you have any doubt.

How To Discover A Real Buried Treasure

The map printed in this book has an honest-to-goodness "buried treasure." A *Walking Liberty*, a pure silver (99.9%) dollar coin, has been hidden somewhere in the St. Petersburg, Florida, area shown by the map. (see Diagram 4) For more fun and easier map dowsing: enlarge the map on a photocopier, or take this book to a fast print shop to have it done.

Now dowse the map with your pendulum as discussed on pages 53 through 55. Discover the "treasure." Mark the location on the map, and send a photocopy to:

Map Dowsing Experiment
Top Of The Mountain Publishing
P.O. Box 2244
Pinellas Park, Florida 34664-2244 U.S.A.

The publisher will send a Commemorative U. S. Silver Waking Liberty coin to *the first five correct answers* sent in, according to the postmark. The Walking Liberty is one of the most beautiful coins ever minted, you may have seen the ads in *TV Guide* running a special sale of the coin for twenty-five dollars. If you would like to know who won the Silver Waking Liberty coins, send a self-addressed stamped envelope to the above address.

You are now off to the world of finding buried treasure, lost items, people, and puppy dogs. If possible, use your "inner talents" daily. Map dowse your questions, your wonderings, your musings. (In Part II, you'll learn how to quantify and get other needed information.)

And do keep a log of your successes.

Your Success Log

Make a Success Log of all your *positive* pendule experiments; that is, your hits. Ignore the early failures (if any), as you will find the more you practice the better you become. The secret is to *focus in on your correct hits* to create a positive frame of mind. It's even a good idea to review your success log before you start to dowse. This helps put your mind in the proper channel or *frequency*.

The Miracle Of Teleradiesthesia

Should You Charge For Your Pendulum Work?

For years we have heard people say you should not charge money for a God-given gift. Tell that to the struggling artist, sculptor, writer, or psychic. Anyone can, *and should*, charge for something they are good at!

Many dowsers don't charge for their services—maybe from a lack of confidence in their talents, or maybe they feel it cost them nothing to dowse. They're forgetting the cost of their most valuable commodity, the one thing they cannot replace—TIME (not to mention ENERGY).

Is it right to charge if you could be wrong? Every day, every minute, every second—someone is charging for information which could be wrong, which could be terribly wrong. Business consultants, stock brokers, public accountants, diagnosticians. Anyone in any profession can be wrong. Some are charging millions for their misinformation!

If you are performing up to the best of your ability, then charge for your time and talent and effort. People value the information they pay for—and they too often feel something for free is of little value.

As you become more proficient with your pendule, and especially if you use your instrument in public (as you should when you need information on the spot), you will be asked by others to dowse for them.

A few will be truly seeking guidance, others will be merely curious to see if your pendulum agrees with the decision they've already made!

If you have a job and family, your energy is limited. Always remember, when you charge for your services you get more sincere clients and fewer time-wasting curiosity-seekers. The decision to charge or not is, of course, up to you.

Tithing To The Person Who Helped You

Although charging may not be approved by some, tithing is a generally accepted tradition in metaphysical and dowsing circles. Tithing is a reward you give to the one who helped you find that hidden treasure, or any other financial windfall. In tithing, you send ten percent of the treasure, sales, or other gain to the one who gave you the information... or to who taught you the system which produced it.

When you hit a really big one from answers your pendule gave you—and you want to follow the tradition—by all means, call or write us in care of this publisher to send us your tithe!

Even if you don't send that ten percent, do let us know about your successes from using the techniques in this book. Perhaps we will put them in our next book. That may be tithe enough.

PART II

PENDULING FOR YOUR HEALTH, WEALTH, AND HAPPINESS

Taming The Wild Pendulum

Solving With The Basic Powell MultiMeter©

The interest in using the pendulum is worldwide, and getting wider. Judi and I have just returned from a tour to Budapest; Athens; and Nicosia, on the island of Cyprus. We taught seminars on "Using the Pendulum for Creative Problem Solving for Business and Personal Success" to literally thousands of people. In Athens alone, the paid attendance for one pendulum seminar was over six hundred people. The participants were especially excited about all the detailed uses of the Basic MultiMeter©. So let's see what it can do.

Just What Is "It"?

You have learned how to make and use the simplest of all psychic meters the pendulum. You have practiced and become fairly proficient in using this "swinging" instrument. Now take the next step to fine-tune your contact with your Higher Self, and learn to *qualify* and *quantify* those simple Yes and No answers.

> NOTE: Using the Basic MultiMeter© should not be attempted until you are pendule-proficient, which means you have positive results with the basic YES and NO process!

You are now going to "attach" the *Basic MultiMeter©* to your pendule. The idea for this new psychic tool came to me (Tag) when I saw a meter used to check electronic/electrical components—that one could change to different voltages by merely switching a dial. It caused me to think of creating a "meter" for switching to different *mind dials*, mind frequencies!

I had two objectives: (1) To design a pendulum meter which would allow more than just Yes and No answers. A meter which could give *percentages* and *numbers*; and *pluses* and *minuses*; and even do *tests* on your body, your car, your cat, whatever! (2) To design a meter which you could switch from one type of reading to another by means of its *Thought Buttons*—hence

Solving With The Basic Powell MultiMeter©

a MultiMeter©. "Basic" because it is the base for your being able to do all this advanced penduling.

Yes, there are other meter designs written about in numerous books on pendules, but few are as "multifaceted" and easy to understand as yours truly! (In addition to the Basic MultiMeter©, I have added a Time Machine, and Judi has created the Compatibility Palette. That makes three Powell meters to use with your pendule.)

You are now about to have the real adventure of penduling, to expand your horizons far and away. Before starting on this adventure, photocopy this Basic MultiMeter© (see Diagram 5) from this book—but only for your personal use. (All three of the *Powell Meter designs are copyrighted and protected by Federal Law*. Permission is granted to you to photocopy them—for your eyes only!)

To make your MultiMeter© more durable, you might wish to seal it in plastic. Most stationery stores will laminate it permanently between two sheets of thermolamination. Or you could buy a roll of clear contact self-adhesive plastic and do it yourself (a full roll would allow one to seal about half-a-daily newspaper). Having laminated it, carry the MultiMeter© plus the pendule around with you in your pocket or purse for a lifetime of years.

BASIC POWELL MULTIMETER

copyright © 1995 Dr. Tag Powell & Dr. Judith Powell

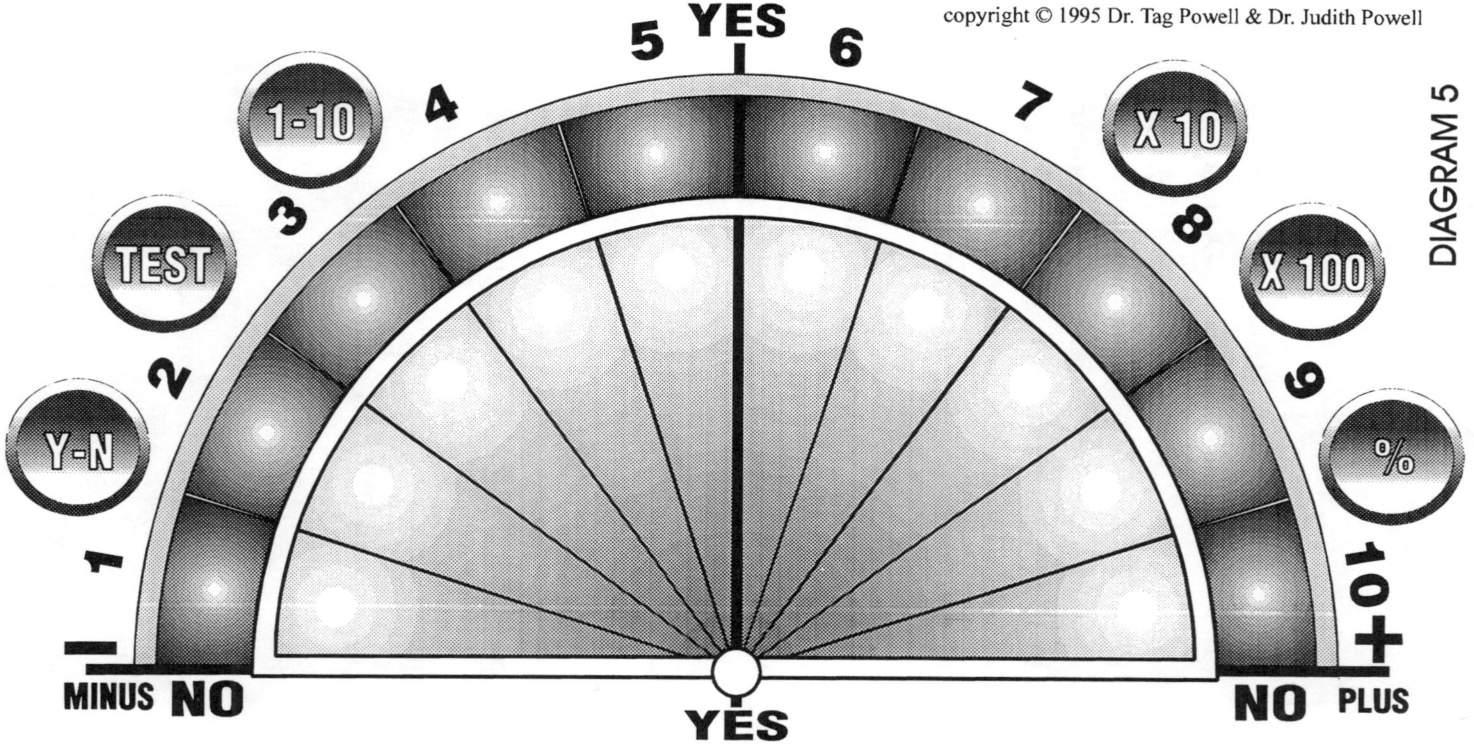

- MINUS NO
- 1
- Y-N
- 2
- TEST
- 3
- 1-10
- 4
- 5 YES
- 6
- 7
- X 10
- 8
- X 100
- 9
- %
- 10 +
- NO PLUS
- YES

DIAGRAM 5

NOTE: *the publisher is now making available all three Powell MultiMeters© in full color and thermolaminated for a small fee (see back of the book).*

Exploring The Powell MultiMeter©

Let's examine the design of this Meter and see just what it can do. Notice at the bottom of the hemisphere the *horizontal line*, that familiar "NO" at the beginning and end of this line.

Notice the *vertical line*: starting at the *top* of the MultiMeter© with the word "YES," down to the *Starting Point* (small center circle), through which your pendule will swing to make its usual Yes path.

Now touch your pendule to the first of your *six Thought Buttons*. Starting with the most familiar, touch the center of the *Y-N Button* (down-left side). This will *dial* or set your MultiMeter© for a YES-NO mode, and

Starting Point enlarged from Basic Powell MultiMeter.©

also set your inner conscious for it. This is not magic. Touching your pendule to the YES-NO Button (or *any* of the other buttons) becomes a visual metaphoric symbol. It helps to define your objective more clearly—in this case, getting a truer Yes or No answer. (see Diagram 5 for the full MultiMeter without health band)

Let's check it out. With the MultiMeter resting on the table or in your lap, touch your pendule on it for a thinking moment. Now lift it up and let it hover just above the *Starting Point*. Ask a question for which you know the answer is going to be "Yes." Note that the pendule will begin to swing up toward the YES word and down below the Meter, giving you a Yes answer.

Now check your No response by asking a definite "No" question. Your pendule should now swing from left to right, following the NO-NO line. Nothing new here, in fact you didn't need the MultiMeter for this. *Always starting your use of it with the YES-NO response*—a response you and the pendulum now know so well—*will anchor your success* in the MultiMeter and thereby increase your accuracy.

Using The MultiMeter© "Test Mode"

Now let's check out the *Test Button*, just above the Y-N Button. It is designed to determine if something—

your diet, your investments, your marriage, your pet, your health—is within an acceptable or unacceptable range.

You can test to see if a specific organ, muscle, joint, or other body part is functioning correctly. You can determine if you are taking too much of a vitamin or too little; if you have the proper amount of nutrients in your system; if you are eating the right foods and if your body is correctly assimilating that food. (But more about "testing" and your *health* in Chapters Six and Seven.)

You can use the Test Mode to determine levels of toxic waste in a stream, river, or lake. Or the toxic levels in your own lawn or garden. If you are a truck farmer, grove or orchard grower, please check!

Also, what a great idea to see if your vegetation is receiving enough or excessive amounts of nutrients. Check the pH level in your soil, the nitrogen level. Check out the "health" of your plants—your azaleas, your rose bushes, your papaya trees! (Use the added health banded Basic MultiMeter in the next chapter, see Diagram 6.)

How To Number Crunch With A "One-To-Ten" Reading!

You are about to do some interesting *number* work. Your MultiMeter can give your pendulum the ability to count; to actually give you a qualitative answer. Maybe you need to plan how many will show up at a banquet; how many boxes to order; how many days to go on a cruise; how many children to have; how many anything.

Or you could ask how much money to charge for a new product; what is your perfect *body weight*; how much to pay for that new car; how much of a raise to ask for; and on and on. Perhaps you should also ask how many times to ask one question. Overdoing it, "nagging" the pendule won't work!

COUNTING PROCEDURE:

You noticed the numbers around the outside of the hemisphere, and wondered what we were going to do with them? To answer your "number" questions, of course! But first, ask the pendulum if your answer *can* be given in a specific amount or size or number of any kind.

If Yes, then touch your pendulum to the *1-10 Button*, the circle at the upper right-hand side of the Meter (see Diagram 5). This sends the message to your inner conscious that you want an answer from one to ten. You have just "dialed" in the Number Mode.

As before, let your pendulum hover above the small Starting Point. Now ask a question that requires an answer within that one-to-ten range (higher numbers we will get to shortly).

Solving With The Basic Powell MultiMeter

Take your time. Since this is new to you, the pendule may take a little longer. It begins to swing; to direct itself toward a *number*. What number is it? Allow plenty of time. It may change its direction after four or five swings, and then settle in and concentrate on one specific number. The right number!

If the number you need is in the 10 to 100 range, touch the *X10 Button*... and think a 0 after each number." The 1 becomes 10, 2 becomes 20, etc. All you did was touch your pendule to the X10 Button, and let your mind do the multiplying. For a number in the 100 to 1000 range, just touch the *X100 Button*, and *think* your basic number times 100. Simple. And amazingly accurate.

If you need a number like 17 or 53 or 102 or 3032—no problem! There are two different ways you can do it:

1. First you get your *base number* by using the *X10 Button*; say it's forty. Now, after touching the 1-10 Button, let the pendule swing again for a single digit. It goes toward 6. Voila! Your answer is 46.

Anything in the hundreds will just take more steps. You get your base number—say, 300

(3 x 100). So concentrate and swing for your double digit by touching your pendule to the X10 Button. It's 20 (2 x 10), therefore 320. Nail it down by swinging for the single number by touching the 1-10 Button. It's 4—your answer is 324.

2. Let's try a quick way when you don't have to be exact.

Get your base number—X100 Button; say it's 600. Now concentrate, and swing, and see what "pie slice" of the MultiMeter© the pendule goes over—say it's in the middle, equaling half, making it about 650. Less than, more than—making it about 635, or 670? Experiment with this. Your answers will get more and more fine-tuned.

Or maybe you want to send out advertising flyers and have *no idea* how many you need. Start with good ol' Yes-No. Do I need more than a 1000? YES. Do I need more than 10,000? NO. Pin down which thousand it is by swinging to the band of single digits and go from there.

By now it's probably obvious to you that there are many ways of doing "the numbers." You'll experiment and find what works best for you. All it takes is your pendule and your psyche!

How To Set Your MultiMeter© For Percentages %

At one *Wild Pendulum* seminar that Judi and I were giving, we had the class do an impromptu exercise about a question we thought we knew the answer to. Boy, were we in for a surprise! Not only that, it turned out to be a great demonstration of the use of percentages!

They were to ask their pendules, "Will Tag's three new books be finished on time for the New York Book Fair?" The Yes-No answers brought a YES for two of the books *Think Wealth* and *Slash Your Mortgage In Half*. When the class asked if the book *Silva Mind Mastery for the '90s* would be ready, the answer came out a resounding NO!

As this was our most popular book topic, it not being ready for the New York Exhibit was a major problem. The next question the group asked was, "Is there anything that can be done to remedy the situation about that third book?" Almost all the pendulums swung to YES.

Now the challenge was to narrow it down, to check on delivery. So, getting out their MultiMeters©, the class now asked, "What were Tag's chances—or *percentage for success—*to have the *Silva Mind Mastery* book to display at the New York Book Fair?"

To find the answer the class touched their pendules to the *% Button*, and each of them concentrated hard on that MultiMeter© Percent Mode (mass concentration can be a powerful thing). Their pendulums began to swing, leveled off, then swung up to the space between 8-9. It meant we had an 80 to 90% chance for delivery!

The next day we called the book printer. They assured us that all three books were okayed for delivery on time for New York. Then we called our account representative, Cheryl Corey, for confirmation. She said she had just returned from the plant and had seen the books being bound.

We were, at first, glad that the pendules had given the wrong answer, but wondered how this was possible. How could so many pendules be so wrong! Our faith in pendulum-dowsing was being sorely tested.

One hour later, we received a phone call from Ms. Corey. The *Silva Mind Mastery for the '90s* book would *not* be ready for the exhibit. They had used the wrong paper for the cover!

The pendules were right after all. Relief. The practice of mind-dowsing was once again validated.

Yet awful—no third book to show in New York! But, hadn't the pendules also said there was an 80-90% chance? So we persisted. We asked Cheryl to find out if there was any way possible we could just have some samples to show at the Book Fair.

Because the mistake was the printer's fault, the company promised to run off a few hand samples, and would have them delivered to us in New York, just in time to exhibit. Wonderful! And, yes, the sample books with the correct cover did arrive on time. A couple of weeks later the shipment with the rest of the corrected copies arrived at our publisher's warehouse. Eventually, this book went on to become our worldwide best seller. It is currently translated into twenty-five languages.

From this "New York story," you can see how the uses of the pendulum can be greatly expanded by your ability to get the *chance* or percentage for your possible success (or failure!)... and how important that can become.

The Percent Mode allows you to check on your chances for successful completion of a project, be it: writing a book, remodeling a house, starting a new business, or the performance of your mutual fund! It also allows you to check on just *how well* a body organ is functioning. What percent of your mind or body health is dependent upon exercise, upon diet. The chances for happiness, fulfillment in a marriage, even the chances of getting, or being, pregnant!

Your pendulum plus Percent Mode can give you what chances you have in being satisfied with a new job, a new location, a new home; a new look, a new group, a new school.

You can even check, as we did, on the percentage of a successful outcome of a business transaction. So just think of "what percent" of your answers can now be improved with the Powell MultiMeter©! It is not only greatly useful in both your business and personal lives, but it's fun and fascinating as well.

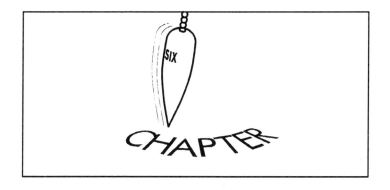

Detecting Illnesses...
Swinging Into Health!

Contacting one's *inner self* has long been a quest for those seeking a *higher awareness*. Yet many of these same people often bypass the Inner Methodologies that can help them, in their *physical everyday world*, to take care of their health and the health of loved ones. Using the inner conscious to maintain a healthy (Godly) physical body should be a concern of all seeking that "awareness," be it higher or inner or worldly.

Tapping the inner conscious, detecting the health condition of a loved one, or even a stranger, miles away, might stretch the envelope of believability for many. Yet investigation into research which has been going on for years could be a mind-expanding experience.

Let's briefly look at the work of our friend Cleve Backster, researcher and scientist, and head of the Backster Institute. You may know of him. His research has been written about in many books and seen in TV documentaries, such as *The Secret Life of Plants*. He is regarded as one of the world's outstanding authorities on the polygraph. Backster has even been called on several occasions by Congress to testify as an expert witness.

In his more than thirty years of research, he has learned that a bond or *communication* exists between the cells of the body—even when the cells are separated from the body, or from each other, by several miles!

In one test sample, cells were taken from a subject, divided, and then moved to two location several miles apart. Both sets of the sample cells were carefully monitored.

Each time an experiment was performed on one cell, a reaction was seen in the "twin cell" miles away. This astounding work can be read about in Backster's (written with Robert Stone) most recent book, *The Secret Life of Your Cells*. It tells how this bond exists in the plant kingdom as well as in the animal kingdom. This *interbonding may even link all members of both kingdoms!*

At the very least, this research may help to explain how our inner conscious can detect problems in another person's body miles away. Or help us accept that there are some people who can understand what an animal is feeling, even "thinking." Or that there are Eastern mystics who can commune with a tree, hear the message of a river.

The Legality Of It All

Backster's testing with the cells demonstrates the fact that *distance is no barrier to the mind.* The barrier to be aware of when "testing" the health of another is a very important law:

LEGAL NOTE: *You may NOT diagnose an illness in the presence of the subject; only a licensed medical doctor is allowed to diagnose. If you are doing any type of health work, your subject or client must NOT be in your presence. The subject can, however, be in a different room or thousands of miles away. There is no law against self-diagnosing.*

We presume the difference is that the AMA does not believe one can *psychically* detect an illness at a distance. Legally that distance, and that subject, can be a thousand miles away or just across the hall. Since *the mind converts all information to electrical energy, time and space are no barriers to success!* So remember to stay within the law, and at a distance, and you can stay out of trouble. You can, however, self-diagnose. People do it all the time. (At their own risk, of course!)

All this means is that we are not here to circumvent standard health advise or to contradict your physician. Our aim is to augment and enhance your bountiful health by you using your innate powers of the psyche.

Testing For Better Health:
Too Much Or Too Little

Now we're getting into some big-time testing! We're going to *dial* in the *Test Mode*. But first, using your Basic MultiMeter© touch your pendule to the Y-N Button. Remember, always do the Y-N check before you proceed to more complex usage of the MultiMeter©—it "dials in" your focus.

Next, let the pendule hover above the *Starting Point*. Ask if conditions are right to accurately use the MultiMeter at this time. If you get a Yes answer, proceed. If it's a No, ask if you are able to change the conditions in order to continue, as described on page 32. Assuming it was a YES, you are ready. Go to it.

Now touch your pendule to the *Test Button* (right above the Y-N Button) and concentrate. You are switching "frequencies," dialing in TESTING. To further reinforce the focusing of your thoughts, write on a piece of paper what you are testing for and also for whom (as long as that person is AWAY from you). Since your pendule will be in your right or dominant hand, you will be holding the paper in your other hand. (Using a piece of paper in this manner is called the *Proxy Method*, as you will see in Allergy Testing in the next chapter.)

Let's say you want to test to find out if you (or a family member) have the proper protein balance. To get a reading on this you will need to "dial" in the *Heath Band*, which is around the edge of your Multimeter. (see Diagram 6)

On this Band you will see two *P*s marking off "pie-slices" on each side of the heavy Vertical Line. The P stands for *Problem*. Above each of these two Ps, you will see on the Health Band two *M*s, marking off more "slices." The M stands for *Marginal*. At the top of the Meter you see S, on each side of the heavy Vertical Line. The S is for *Satisfactory*.

Using The Health Band

Let's try your question using this Health Band. Dial in the Mode by touching the pendule to the *Test*

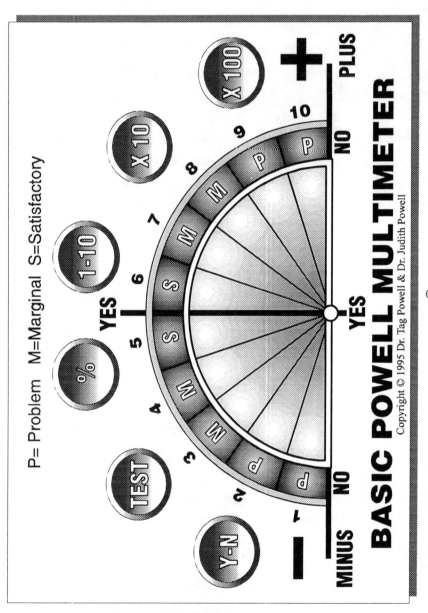

DIAGRAM 6

Detecting Illnesses... Swinging Into Health!

Button, and *think Health.* Now let your pendulum hover over the Starting Point. Concentrate, concentrate. Ask, "Am I ingesting the correct amount of protein for my body to reach or maintain vibrant health?"

For both the M and the P ranges, your pendule will swing on a *diagonal*; the M swinging higher and more toward the S than does the P, which swings much lower, almost to the NO-NO line. But if it's S, it will swing the good ol' Yes line, into the top pie slices.

If your pendule responds in the S range for *Satisfactory,* you are eating the correct amount for your body at this time. But if it swings in the P range—you've got a *Problem.* On the other hand, if it swings into the M range, you are *Marginal* and had better take steps to bring it up to S.

It's also very important to note if your answer was Problem or Marginal, which *side* of the Vertical Line did your pendulum chose to swing on! If it swung on the PLUS side, you're getting *too much* protein (or too much of anything for the health of whatever). And, naturally, if it swung into a "slice" on the left or MINUS side, you're getting *too little.*

Too little is rare indeed in beef-rich, protein-gobbling America! But should you get this answer, you may have a *serious* protein shortage and should

87

contact your health caretaker. (Are you undertaking being a "vegan"? A TOTAL vegetarian? To be a vegan is fine, if you have full knowledge of just how to healthily be one—otherwise, big problems! DO go about it with proper guidance from a nutrition expert!!)

But if you're like too many of us and getting too much of a "good thing," going overboard on protein can lead to high cholesterol, high blood pressure, clogged arteries, overloaded kidneys, etcetera. Again, check it out with your health caretaker. Either way, too much or too little, is a "problem"!

Testing, And More Testing

With the Test Mode and the Health Band you can check out the functioning of any organ or part of body. Let's say you want to check John's liver (or Susie's kidneys, or Mom's knee, or Dad's stomach, or Uncle Henry's gallstones.)

On a piece of paper write "John" and "liver." Holding the paper in your left hand, touch your pendule to the Test Button. Concentrate on John's liver. (Whenever you want to test anything—liver, car, cat food, whatever—always first touch the pendulum to the Test Button. Set your mind and "dial" in the Test Mode.)

Now let your pendule hover over the Starting Point and state clearly and firmly, "Give me the functioning of John's liver." Does the pendulum swing into the S for Satisfactory, or into the P area, or M area? Write down the answer. You could further ask, "Is there anything I can do to help?" (Is there anything John can do to help himself?)

Do always remember to check anyone for the possibility of bulimia, anorexia, alcohol/drug abuse, pneumonia, cancer, TB, immune deficiencies, or anything else you think is pertinent.

Don't be content just asking about the physical health. Always be thorough enough to inquire about that person's emotional as well as spiritual health. Such problems are not often discussed among the family, and sometimes not even recognized by the individual. We now realize one's emotional health can affect, for good or ill, the *physical* body.

Later, some gentle questioning or a few well-chosen words from you to that person may be of great help. Most people appreciate, respond to a sincere interest in their well-being. But don't pry.

> HEALTH NOTE: *If you find you need a large increase or decrease of anything you are currently ingesting or doing, you should check with your health caretaker before making any drastic changes in your prescription, diet, or activity.*

Penduling With Body Illustrations

By using the following *Body Illustrations* plus your MultiMeter© *Test Button*, you can do an overall check of yourself, friends, and family.

You will also need a pen or pencil *pointer*, and a *pad* for notes. You might want to make enlarged photocopies of the Body Illustrations (see Diagrams 7 -14) so they will be more usable.

We're giving you here just a few very basic body illustrations. If you run into something you want more details on; let's say, a possible nerve disorder, a circulation problem, a pulmonary condition, we suggest you go to a good medical dictionary, one with plenty of illustrations. And then, of course, to your doctor!

Heath-Check Questions

First ask, as always, if conditions are right for testing.

If yes, follow that with: "I am checking the health of," (write your name or that of test-subject on the pad) "for existing, or possible near-future problems." Now you are ready to begin checking the body. (This could be a good way for a worried wife to see if her husband's pain is indigestion, peptic ulcer, hiatal hernia, or heart!)

Detecting Illnesses... Swinging Into Health!

1. Starting at the top, ask if there is a problem in the subject's *Head Area*. Place the pointer in the approximate center of that area on the Body Illustrations. (see Diagram 11)

If the answer is NO, proceed to the Upper Body Area.

2. But, if the answer is YES, while still in the Head Area, touch your pointer to the *Brain*, and ask: is the problem in this area? If NO, proceed to the face. Ask about the eyes, ears, nose, cheeks, mouth, jaw. Don't forget the teeth! While checking the nose, perhaps you get a YES. Is it sinus? NO. Could it be a deviated septum? YES! (So that's why the snoring!) *Note that on your pad.*

3. You now proceed to the *Upper Body Area* which includes neck, shoulders, arms, hands. Is there a problem in the throat area? YES. Is it a problem with the tonsils? NO. A sore throat? NO. Is the problem glandular? YES. Thyroid? YES. Pin it down. Hypothyroid? NO. Hyperthyroid? YES. *Write that on your pad.* And always, when you get a YES, double-check it, at two different times, and then consult your physician!

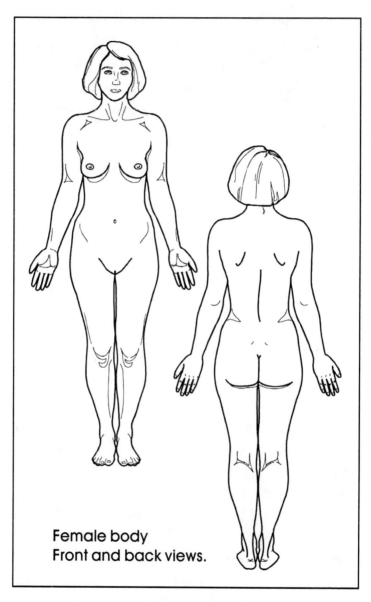

DIAGRAM 7

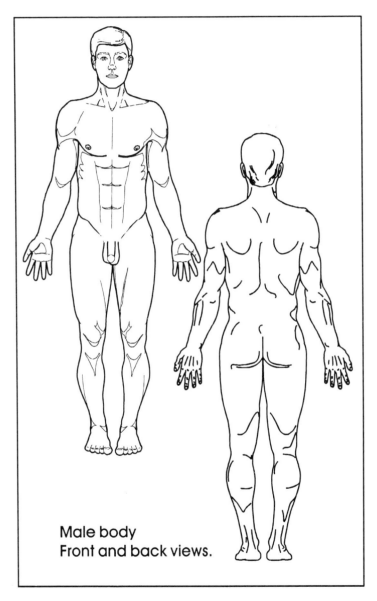

Male body
Front and back views.

DIAGRAM 8

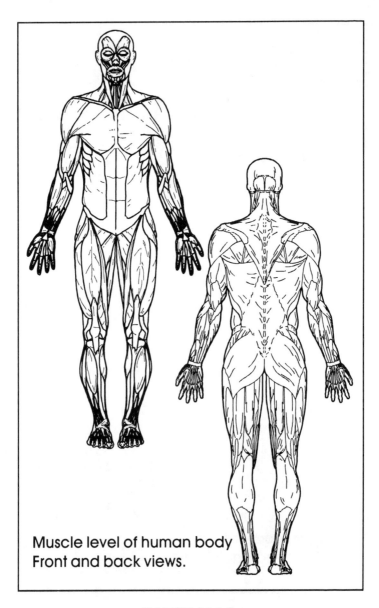

Muscle level of human body Front and back views.

DIAGRAM 9

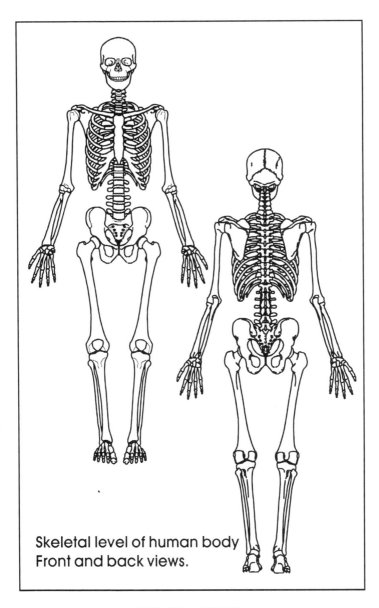

Skeletal level of human body Front and back views.

DIAGRAM 10

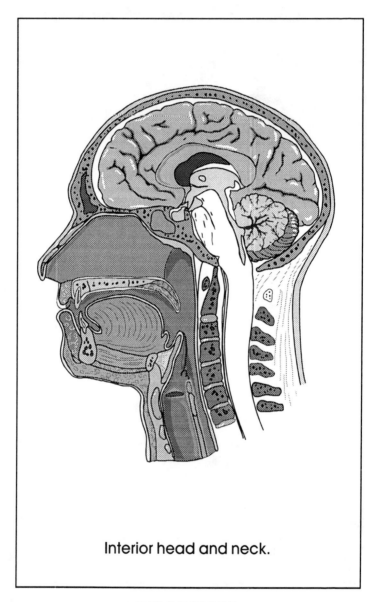

Interior head and neck.

DIAGRAM 11

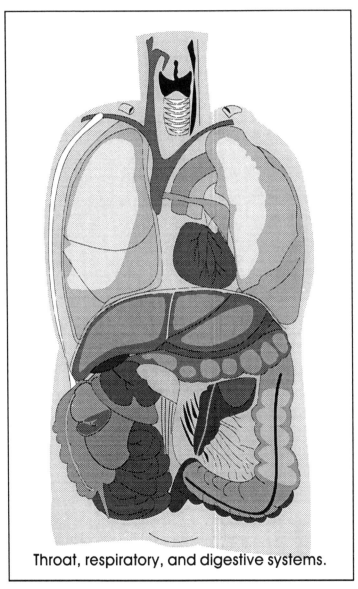

Throat, respiratory, and digestive systems.

DIAGRAM 12

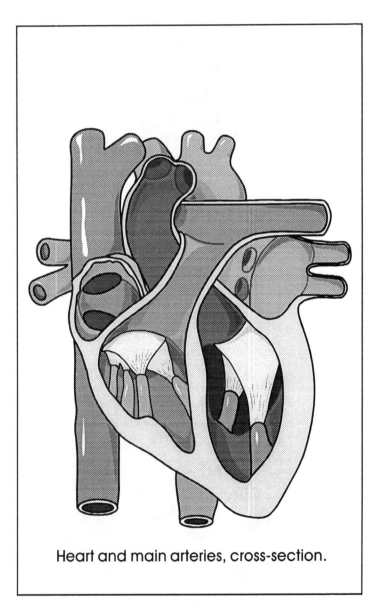
Heart and main arteries, cross-section.

DIAGRAM 13

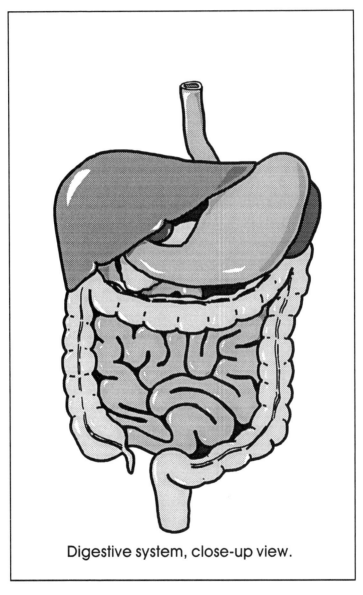
Digestive system, close-up view.

DIAGRAM 14

4. Now do the *Shoulder Area*, continuing down through upper arms, elbows, forearms, wrists, hands, fingers—remember those finger joints. You will, of course, be checking not just bones—but joints, muscles, tendons, skin; organs, veins, arteries, nerves, and possibly an ingrown toenail! (see Diagrams 7-10; Diagrams 12-14 for internal organs)

5. After you have gone over the entire body, from head to little toes—just following the Body Illustrations and the above questioning procedure—you should then go back to your MultiMeter©. Wherever you noted on your pad there was a health challenge, check out whether Marginal or Problem, using your pendule and the Health Band. (see Diagram 6)

What all this amounts to is a complete physical done with the psyche. A *psychic physical!*

You owe your loved ones a frequent pendule health-check, whether of mind or body. And this goes for you too. Check *yourself* out. Consider doing a psychic checkup at least three or four times a year. So if a problem is in the making, in you or "them," you will be aware of it before it has a chance to become critical.

If we seem to be writing a lot about health, it's because we think it's that important. It's hard to go for wealth or happiness if you or they feel unwell.

Taming The Wild Pendulum

Penduling For Allergies

The Perfect Body has no headaches, tummy-aches or allergies. If these ailments afflict the body, something has caused an imbalance in the physical system. What you take in for body fuel, your food, can be a major cause of body malfunction. Most of us are aware by now that some foods can cause major disturbances in some individuals, and not in others.

Sometimes a health problem is caused by an ALLERGY to a food—and you don't even know it! The solution is so simple: use your pendulum to discover if you DO have any allergies, and to what specific foods or substances.

YOUR FIRST STEP: assemble food samples for testing. Since it is believed that "you are what you eat," test the foods you eat daily. Always remember that as a beginning pendulist/dowser, you should flavor your answers with a grain of psychic salt!

For example: before drinking a glass of water, you ask if you should drink the water and your pendule may say NO. Now you know that water is necessary for health and survival, so the answer could be wrong (it happens). Of course, there's the possibility you should not drink that specific water, or from that glass, or at that moment.

Three Testing Techniques For Foods And Liquids

You have "warmed up" your psyche by testing your Y-N Mode with a known question. Then you "dialed" in your Test Mode by touching the Test Button with your pendulum, and mentally focused on just what you are testing for. Now you and your inner conscious are ready.

There are basically three schools of thought about what you need in order to test samples (of food and much more).

SCHOOL OF THOUGHT #1:

The dowser must use actual samples of the food. For this model you have five choices of action:

Penduling For Allergies

A. Holding the sample in the palm of your left hand, penduling over using the opposite hand, swinging YES or NO.

B. Holding your palm, face down, one or two inches over a sample on the table... pendule in opposite hand, swinging to choose one of two small pieces of paper, with YES written face down on one, NO on the other. (see "Judi's Solution—Don't Peek!" on page 45).

C. Placing the sample in your mouth, allowing it to assimilate, then letting the pendule choose a YES or NO paper.

D. Putting the sample on the inside of your arm at the left elbow, folding the arm upward. Letting the sample be absorbed through the skin into your system, while letting the pendule choose a YES or NO paper.

E. Or simplest of all, especially if you're dining out and you're not concerned with looking like you're from Outer Space (but what a great conversation piece!)... letting your pendulum swing over a particular food. Ask, "Am I allergic to this food sample?"

Wait patiently for all your answers. This is a really new kind of "testing" for you and your pendule and it gets even newer.

SCHOOL OF THOUGHT #2:

The dowser can use a Proxy or Symbolic sample of the food to be tested.

Your *proxy* can be the *name of* the food written on a piece of paper, or a *picture of* it. Remember, in Map Dowsing in Chapter Four, the term *"witness"*? That whenever something is not the real thing—is a *proxy* or symbol—it's called a "witness." Here, your witness for the real thing, the food sample, is the "name" or "picture." This can be carried to marvelous extremes; such as using a name/picture of a car, house, boat, cat, dog, canary. The sky's the limit when you're dowsing with the *Proxy Method!* Now let's do it.

A. Holding the written name or picture of the food in your open palm... letting the pendule in your other hand swing YES or NO over it.

B. Holding your palm face down over the paper on the table (*feel* the distance that seems truest to you)... letting the pendule swing YES or NO.

C. Pointing to the paper on the table with a pen, pencil, or your finger... letting the pendule swing YES or NO.

SCHOOL OF THOUGHT #3:

The dowser need only THINK of the food for testing.

A. Thinking about, or *visualizing* the sample while holding the pendule. (Neither sample nor proxy required; *it's all in your mind).*

Which one of these methods is the best? Although many swear that one method in particular is the only way to get the "true" reading, it is probably only a matter of *belief structure*. Whichever one seems most feasible to you will probably reap the best results. We know each will work, different ones at different times, for we have successfully used all these methods.

For some, School of Thought #3 may get the weakest results—as it may seem easier to focus on an actual object, or at least a written symbol or picture. Others, who are more into *visualization techniques*, will appreciate the convenience!

Also, the use of the *actual samples* could have certain drawbacks. For instance, in restaurants you've brought your "portable testing tool" to check out just what it is you will be eating and just what effect it will have on you. Be aware. These real-life samples are probably cooked in oils and spices which may cause a false reading—yet the food itself, when checked without additives, may register a healthy NO: you are NOT allergic! To get around this, ask very SPECIFIC questions.

Questions, Questions, And More Questions

Since the wrong question can produce a wrong answer, when you form your questions be thoughtful and thorough. In the preceding paragraph if you got a "Yes, you are allergic to this food," be sure to follow up with, "Am I allergic to the oil the food is cooked in?" If No, ask about the spices or seasonings. Ask about chemical additives (MSG!).

It's a good idea to begin with: "Does this food have something added that might confuse the test?" If not, then ask if the food by itself could cause a problem in your body that would lead to an allergic reaction, or worse. If the answer is Yes, consider eliminating that food from your diet.

But be careful of the actual phrasing of your question. Asking, "Is this food bad for me?" is too nonspecific. *What* is bad and how bad is bad? Probably the only food that is really good for you is organically grown, with no chemicals, pesticides, or steroids; or without gassing them or fooling with their genes. Ask your pendule about all this. Later.

For the next three days, use your pendulum to test anything you put in your mouth: foods, vitamins, minerals, liquids. (You might even check out the scotch you're drinking, and then a "healthy-for-your-heart" glass of Bordeaux!)

Trusting your pendule and your inner self, you are about to eliminate from your diet those items which test an allergic YES. Having done so, wait a week or two. Notice if you can tell any difference in your stamina, your energy level.

Also, if it seems you are allergic to several items, only eliminate one at a time. Let your body (and mind) get accustomed to the lack of it.

Allergies Are Not Just To Food!

How many people do you know who start convulsively sneezing the minute they walk into a house where there are cats (or where cats were years ago)? Or pediatricians who state that a baby or small child could get a lifetime allergy problem from being around a pet? For the child's sake, as well as yours, by all means take your pendule to the pet shop. Better than being sorry later. Why not also check out your pet's allergies!

What about allergies brought on for the sake of beauty? Women who wear contact lenses should *not* wear eye liner, especially under their eyes. It could do serious damage; and you might not realize it until you are in real trouble! So take your pendule to the drug store when buying any new cosmetic—and remember to pendule-check your "drawer full" at home. Something you may have worn for years could be having a slight but nevertheless real effect on your health. And DO check your prescriptions. Then off to your doctor to double-check if a medication(s) could be causing a problem. Don't forget to check those over-the-counter pills.

How about allergies to certain plants? The sap of allemandes and oleanders. The very toxic sap of the jatropha bush, a popular plant in the southeastern United States. And often the fruit, particularly the pulpy inside skin, of mangoes. Of course, asthmatics by now know to stay away from punk trees!

So when you go to a nursery to buy a pretty plant, an elegant shade tree, a blooming fruit tree—take your pendule! Check them out. Are they for you? You might even hold your pendule over a bag of seeds. Little ones can grow to be big offenders!

Also, pendule the building you work in. Some people working in air conditioned buildings, especially in hot and humid climates, are beginning to develop new allergies to mold, to mildew.

By now YOU will be thinking of all the other things you can test with your pendulum. All the marvelous ways it can serve you. Your most convenient, most marvelous "servant"-tool-instrument.

Bobbing With Your Pendulum-On-A-Stick!

"Great fun..." "wonderfully useful..." are words often heard describing *the Bobber, the happy puppy of the dowser*. Many state that the Y-rods (the forked stick of traditional dowsing), and the L-rods (see Diagram 1) are capable of moving with great strength and precision—but that the bobber seems to have a mind of its own.

The problem with the bobber is keeping it steady (and disciplined!). Most of the time it seems to just want to bob up and down and jump around. That's why the bobber seems like a happy puppy; wagging its tail and following your directions, sometimes, but most of the time doing its own thing.

What today we call the bobber was possibly the "wand" of the *Old Testament* and probably the instrument used by Moses to strike the stone and bring forth water.

So what's a "bobber"? It's a flexible shaft or rod, generally three-to-four feet in length. The original was probably just a reasonably straight long branch or stick. You can use almost anything that's flexible: an old cane pole or fishing rod or even an old car radio antenna. Pick one up at a junk yard and you've got a convenient-to-carry, collapsible bobber!

The bobber is *ideal* for anyone who has difficulty using the smaller, lighter-weight pendule or who doesn't have a steady hand. This puppy *pendulum-on-a-stick* can be the "answer," and will happily answer all your questions.

How Do You Use It?

One end of the bobber is to be held in the hand at waist-height, stretching out the heavier end horizontal to the ground (see Diagram 15). When a question is asked, the weighted end *bobs up and down* for YES, and *bobs right and left* for NO. (It knows the universal Y-N code, just like your pendulum!) When it really gets going, the bobber will whip about with great strength. Other times it just bobs around in a

DIAGRAM 15

circle... when it is not sure, or needs another question asked... or when it has nothing else to do (the happy puppy mode!).

The bobber bobs better if the end stretching out away from you weighs more than the end being held. Holding the smaller, lighter end lets the taper of the shaft arc down and "bob" up, down, and around with the slightest motion. But remember to hold the small narrow end to let the heavier end do the bobbing.

> *SIDE NOTE: Although we believe a flexible stick with a bit of weight on the end will answer all your bobbing needs, the absolute ultimate bobber is a complicated two-spring-loaded devise called the Aura Meter. It was developed by Master Dowser Vern Cameron, and can be purchased from The American Society of Dowsers, for about a hundred dollars.*

Training Your "Happy Puppy"

"Training" the bobber is simple. Just follow the instructions in the second chapter, *Training Your Wild Pendulum*. As always, play twenty questions. Not just, "Am I near the lost treasure of..." but anything you would ask your pendule, like, "Am I going to give a successful party Saturday night?" Always be patient, wait for the answer. Allow it to settle in before you decide it's a definite Yes or No reading. In the beginning, especially, it may play the "happy puppy" game. The more you use your bobber, the more you'll get used to it and its play, and the sooner you will get true answers.

Practice "walking the bobber," just walking along, letting it do its thing. The bobbing movement will make your *bobber-dobber*, as Judi likes to call it, seem like the happy puppy—until you learn to read its motion. Its bobbing up and down YES will help you eventually locate "the spot," or will tell you your true answer. As with all instruments and all skills... practice... practice. The ideas you'll come up with for "exercising your bobber" will become almost endless!

Creating Your Bobber-Dobber

A couple of years ago, I (Tag) was asked to write a series of articles for *New Awareness Magazine* on various dowsing instruments. As a believer in the tradition of "don't talk the talk if you don't walk the walk," I thought it might be a good idea to *make a bobber*. I wanted to do some testing for this article, instead of just talking about all the bobbers I have known and used through the years!

I decided to make it out of a brass-coated, mild steel, gas welding rod. I found this rod material in a perfect three-foot length at welding supply shop—at a cost of less than twenty-five cents! Not sure of the right width, I bought two: $1/6$ inch and $1/8$ inch.

My next stop was the fishing department at *Kmart* for weights (egg-shaped lead sinkers) to put on the extended end of my bobber. (Welding rods are without taper and therefore without a heavier end.) Wanting to experiment with different weights, I bought a box of assorted "lead sinkers" ($2.89). Since I like things that look nice, I wanted a fancy handle. This took a couple of stops until I found a hardware store which carried old-fashioned wooden handles ($1.49). Just perfect to finish off the two bobbers.

Experimenting With Your Bobber

After some experimenting, I learned that both widths of welding rods worked great—the $1/6"$ one seemed to have a bit more swing, but all in all, I liked the feel of both (Judi likes the heavier sinker). For the $1/6$ inch rod, the number 10 lead sinker worked best; for the $1/8$ inch rod, a number 7 lead sinker did the job.

By making a slight bend with a pair of pliers, about a quarter-inch from the tip, I was able to wedge the sinker in place. This seemed to hold fine, although I thought to later place a bit of epoxy for sturdy permanence. (see Diagram 16)

Bobbing With Your Pendulum-On-A-Stick!

At that point in the constructing, Judi arrived. After carefully listening to the explanation of my "major" investment of $2.89 for fishing sinkers when I haven't fished in years, Judi said doubtfully, "Okay, let's see these *dobbers* bob."

Five minutes later she was using the bobber like a pro. "I love it! Why haven't you made me *a dobber* before?"—She mixed *dowsing* and *bobber* and got *dobber*. Then she burst out, chirping, "Wow! Holding this *dobber* I can actually feel a tingly sensation right down my arms to my fingertips! It's like a force of energy flowing out of my hands... like an electrical surge! It's my favorite!"

For a moment I thought Judi and the bobber were going to take off for Bobberland! Bobbing off into the sunset, never to be seen again.

She spent the next hour joyously playing with this "magic wand," demanding and getting answers to one question after another. She finally said, "Okay, make one for the front office, one for the back office and two for home, and don't forget the car!"

Then Judi asked me, "Do you think it's because I'm a happy puppy too, and that's why it wags for me?" She continued, "For ME it swings in 12- to 18-inch arcs, and just bounces its Yes or No! When YOU use the *dobber*, it only moves in minute movements, more like

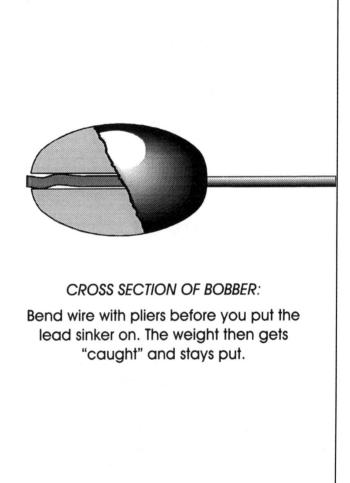

CROSS SECTION OF BOBBER:

Bend wire with pliers before you put the lead sinker on. The weight then gets "caught" and stays put.

DIAGRAM 16

Bobbing With Your Pendulum-On-A-Stick!

the pendulum. Could it be because of my 'taking' to it with such enthusiasm that it has 'taken' to me?"

My wife loves to personalize everything. So I started calling the bobber her "happy puppy." She still loves her "bobber-dobber," as she *insists* on calling it. I must admit, I had forgotten how much fun the bobber could be.

So what are you going to use when searching for water, gold, oil, uranium, or for land-dowsing in general? Y-rods and L-rods. But if you have "the touch," as Judi certainly has, there's just no telling what you can come up with. If it's lost keys, a sales contract, or lunch—the bobber does it! If it's a burning question, a "shall I" or "shan't I"—the bobber does it. Anything a pendule can do, the bobber can do—for it's just a *pendulum-on-a-stick!*

You may or may not get the same sensation... the tingling that Judi felt (still feels!). Or the bobbing and the bouncing. But do try the bobber and see if it LIKES you! And just go bob-bob-bobbin' along... like the Red Red Robin.

Taming The Wild Pendulum

Bobbing For Lotto With Judi

Are you like a lot of us, eagerly dreaming of *winning* the lotto jackpot? Waiting for the sky to open and rain not pennies from heaven but millions!

Do you wonder why it is that some people win big after playing several or just one time and others who play lotto all the time don't win a thing? It is especially unnerving when that one-timer wins, a guy/gal who never played before. Or someone who is so doddering and tottering that it makes you bleat, "But I can enjoy it so much more!" Is it fate, kismet, "the luck of the draw," or just being at the right place at the right time? Let's see what we can do about that.

To increase your chances of getting what you want is to first firmly *know what you want*. So, before you begin bobbing for lotto, sit down with pen and paper in hand and write SPECIFICALLY what your plans are for this great amount of money *when* (not if) you receive it. (The model for a carefully thought-out, target-type statement is described in detail in our *Silva Mind Mastery for the '90s* book.)

You're probably thinking, "Yeah, of course I know what I want... a new house, a new car a ticket for around-the-world cruise." But, have you ever written all this down? With specifics of architectural house plans, a particular car model, a definite cruise ship—plus how much each will cost, etc.? So once you BUY all that you want, what will you do to BE what you want after the "toys" no longer amuse you?

The moral of this story is: *Get your mind in gear so you can handle wealth*, because your inner self knows if lots of money will destroy you or those around you. If that is a possibility, your inner self will keep you far from it, and no amount of dowsing will bring you closer to your money goal! 'Nough said. Now let's move on to just how you can become luckier, assuming now you can handle it!

Bobbing For Lotto With Judi

How To "Bob" For Winning Lotto Numbers

STEP 1: Take same-sized squares of opaque paper (business cards work wonderfully) and write the numbers 1 through 49 (or whatever the "number spread" is in your locale). Turn all the cards over so you can't read the numbers, and thoroughly mix them up. Once they are mixed, spread them out on the kitchen counter, table, bed, floor, or any area large enough to fit all of them. Space the cards about an inch or so apart. Now you're ready to *"bob-dob"* for lotto.

> NOTE: *Since we're talking big money here, and making big changes in your thinking about money, it is of extreme importance that you RELAX. Take a deep breath, let it out to clear your mind and being of negativity. Take another deep breath and wait a couple of minutes before you bob for your winning numbers. (Always dowse in your "connected" inner space, never in strung-out dire straits!)*

STEP 2: With your *bobber-dobber* in hand and arm outstretched, touch the tip of the bobber to the first card/paper and ask aloud:

"Is this one of the WINNING LOTTO JACKPOT numbers (state definitively 'between 1 through 49,' or '1 through 9,' or whatever the number spread), for (say date), in the state of (say state/country), for $ (say dollar amount)?" (now wait for answer)

In other words, you are being very specific to key in accurate and specific numbers. Remember to put heartfelt *energy* into your asking!

If the answer is YES, move that card forward from the rest. If the answer is NO, leave the card there. Remember, NO PEEKING until you have completed the entire series!

Continue on down the line, touching all 49 or however many cards. When you have the number of YES cards you need, NOW you can turn them over. Pencil in your "winning" numbers and go place your bet!

STEP 3: If your bobber-dobber responds YES to *too few* numbers, dowse the leftover numbers again.

If your bobber-dobber responds YES to *too many* numbers, clear the area of all the NO responses (don't peek at them). Line up all those too-many Yes cards, and repeat Step 2 until you get just the right amount of Yes cards.

Tag and I find that we are more accurate if we dowse *only once*, rather than dowse for that same week's jackpot day after day, time after time. It's almost as if your mind is saying, "Oh, you don't like the numbers I gave you? Well, how about these?" Then to appease you, your mind will respond "Yes" to perhaps some of your favorite numbers, rather than the future winning numbers! (I get a headache if I bob-dob the same thing over and over again. I'll feel definitely confused, and that's no way to win, at anything! Remember, *confusion is the opposite of goal-oriented direction!)*

What you're doing when you insist on bob-dobbing again and again is just negating your original dowse—and you've probably got enough negativity about winning, and money, and life! We're trying to change all that, aren't we? So just put some *good energy* into it once a week. Use your common sense!

Remember to always use your intuition. *Feel* the numbers. Feel winning! Make a game out of it, stay "flexible," and above all else HAVE FUN!

Be sure to save those numbered squares of paper to use again for next week's lotto jackpot.

I should also mention, there is a school of thought that states: once you have bobbed/picked/felt your special numbers, you should stick with them week after week. Others feel a fresh "draw" is best. Dowse and you decide.

Using Your Pendule For Lotto

Since I enjoy the bobber-dobber (that pendulum-on-a-stick), I continue to use this method. However, if you wish, you can use your pendulum to dowse for lotto.

Of course whatever you do for *weekly Lotto* you can do for *daily Lottery!* Just remember to know AND say which you mean when penduling... either lottO or lottERY... and play which you mean.

STEP 1: Lay on a flat surface the actual lotto form you would normally pencil in. Now take a pointer (your pencil) and touch it to the tiny Number 1 square.

STEP 2: Ask aloud very firmly, "Is this one of the WINNING LOTTO JACKPOT numbers (state 'between 1 through 49,' or '1 through 9,' or whatever), for (say date) in the state of (say state/country), for $ (say dollar amount)?" (wait for answer)

Continue to move in the right direction (the correct number order) until you have received a YES response to the wanted six or seven numbers (or however many you need for that locale). Pencil in your "winning numbers" and go place your bet!

For this first-time run, only bet on ONE ticket or $1.00! To bet more than one ticket on the same numbers is a waste of money. It only takes one ticket to win! To bet on several different series of numbers is like telling your inner conscious you don't trust the original set of numbers you were given!

Of course, you don't have to bet at all until you feel more certain. You can dowse for the numbers, and then just watch the balls bounce on TV to see how accurate you are. One drawback: If you don't bet and all your numbers are winners, you'll just kick yourself all the way to the moon!

So far, I have bobbed for the lotto jackpot approximately ten times, and have been accurate on up to four of the six numbers. A better percentage than pure chance! I'm not a big gambler, so maybe I'm not winning big because my heart isn't wholeheartedly in it. Also, I usually forget to do it unless something reminds me (like a jackpot up around 20 million!). But those few times of hitting convinces me that *dowsing for dollars* and penduling for winning lotto numbers does work!

(Please send your success stories of over $100 winnings to the publisher. They will give them to us, and we can inspire people attending our "Wild Pendulum" seminars with your BIG WIN. We "thank you" in advance.)

> *TIP: To practice and build up your confidence in your abilities, you can pendule for getting the three, four, or five consecutive DAILY lottery numbers.*

Use your Basic MultiMeter with the band of numbers 1 through 9. Simply ask, "Which is the *first* number of the winning "play four" lottery for (say date), in the state of (say state/country)," (wait for answer). Without betting any money, check the TV to see how accurate you are. This way you can be sure you are "reading" your bobber-dobber correctly.

Other Places To Increase Your Luck

Carry your luck with you. Take your pendule (the larger "bobber" may get you thrown out of the casino!) to Las Vegas, Reno, Atlantic City, or up the Mississippi; to Hong Kong, Monte Carlo, the Bahamas. Check out which one-arm-bandit is "ready to give up." When you get a YES, play that bandit! Once you've quickly scored, ask the pendulum if you should stay on that one. If NO, move on and dowse the gambling floor for another "ripe" machine! You can also dowse the blackjack and roulette tables.

To Gamble Or Not Gamble

We're not necessarily condoning gambling, but if you do gamble, *you* might as well be the one to "stack the odds"! Now think of other ways to increase your odds, your chances in life, to win with your wonderful valuable dowsing tools: the pendulum and the bobber-dobber.

In essence, when you do this "bobbing for lotto" you are in reality bobbing into your future; that is penduling *today* what those ping-pong balls will read *tomorrow* or a week from now. Try it, and watch bobber-dobbing increase your future odds of winning.

When you win at one thing, you'll start winning at many things. The winning goes to your head!—to your psyche, and you begin to feel, act like, BE a winner!

For you to stay in good with LADY LUCK remember to always spread your future wealth around—don't hoard it! Don't waste it. Use it!

We wish you Good Bobbing!!

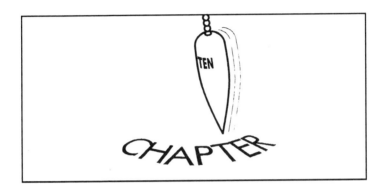

Discover Your Perfect Mate, Friend, Pet, Or Business Partner!

We've done Health, we've done Wealth, now let's do Happiness. For without that, the first two just don't make it. "It *don't* mean a thing if you *ain't* got that zing!"

To work up to happiness, let's check back into the metaphysical community. Here we often see people who "talk-the-talk" but don't "walk-the-walk." People who talk a good game about psychic abilities but don't use their God-given talents to make their own lives easier, better, happier.

Before you resent my (Judi's) finger-pointing at you, know that three fingers are pointing back at me. For as much as I try to stay on the pathway, I can still find myself enmeshed in the everyday business world. Several times, when my inner-conscious mind has told me not to, I have hired people who just didn't work out, didn't "fit." Metaphysics or not, it is often hard for me to accept that people are what they are. That I (or you) cannot change them one iota. I am reminded again and again that each of us can make only our own reality—and not that of anyone else!

Too many of us go through life making decisions for or about other people, expecting them to think or act in the way we think they should think or act. Let's just take care of our own "walk," for no one can or should try to change anyone else. They have to do that. Maybe that's why we are all on this plane—to do the best we can, at growing and changing ourselves, to be more of what only *we* can be.

One of the most obvious examples of wanting to change, mold, reform someone we can see every day, all around us, in the *mating game*. Just look at people, perhaps even you, searching for *the other*. We meet someone who seems "just right"—and then we try to change them, to make them even more right! More like us. Or more like our image, our fantasy of the ideal mate. (You know that marvelous story about the man looking for the perfect woman... and then he found her. But she was looking for the perfect man!)

We leap into love (into friendship, into business, into, into). Then pain-filled weeks, months, even years later, we realize that person wasn't right for us after all. Crash—into disappointment, disillusionment.

How about the employer who expected something other than what the employee had in mind, or what he/she had in ability? Or when looking for a business partner or investor, do we too often measure that prospective relationship by the other's available funds? Contacts? We forget that a partnership is, in a true sense, *a marriage in business.*

Be it marriage, love, business, friendship—the key word for all successful relationships is *compatibility. When there is true compatibility there is little need to change that other person.* No false dreams exist on either side.

The kind of "compatibility" we're talking about is not in the wishing-heart's eye or the profit line. It's not in the clouded mirror. So how do we "clear the mirror"? We TEST for a real kind of COMPATIBILITY!

By using my unique Compatibility Palette *"Compat Palette,"* you have an easy way to see if

YOU are "walking the same walk" with a NEW friend or lover, business partner or investor, employee or employer, landlord or tenant! Or a perfect vet for your pet. Or a prospective baby-sitter for that most important being in your life. Test your compatibility (your CHILD's compatibility) with a housekeeper, a nanny, a day-care center!

Test for a trustworthy house-sitter. Test for a potential remodeler\contractor. Test for "compat" with your investment broker! A dentist, a doctor; a teacher, a school; a roommate, a lab partner. Test to see if you are compatible with your chosen school major, your present career. The possibilities are as endless as your imagination.

You can even test with a prospective product or service. (Yes! You can use it for nonliving, mechanical, *electronical* things too. Are you compatible with your VCR or with your computer?)

And why not check for Compat with PLACE? Where you live, where you think you want to live, where your company is relocating. What about your dream vacation home, your future retirement home? On and on. Your compatibility with anything. Life itself. (See Diagram 17)

With the Compatibility Palette, you will now be able to use more of your higher intelligence, not your feelings, to be able to make the right decisions.

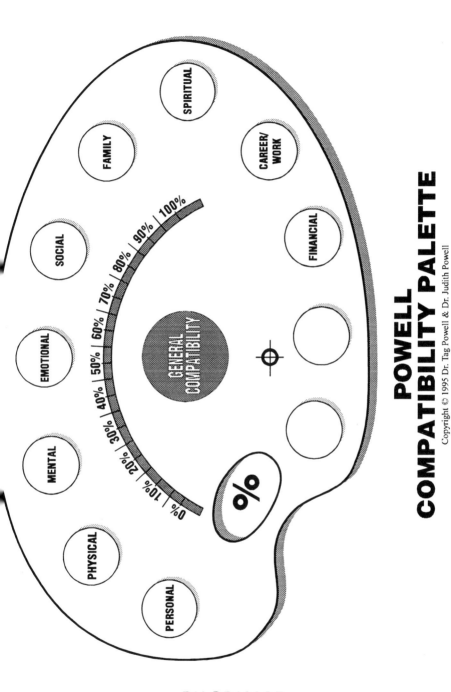

DIAGRAM 17

Looking for that special person, you'll have an inner perspective aiding you, rather than rosy glasses or Barbie Doll dreams, or country clubs, or dollar signs dancing, blinding you.

Yes, of course you can do a fast "rough" test by dowsing for a YES or NO answer. However, using my Compat Palette is a thorough "fine-tuned" way to check everything you just might overlook in your enthusiasm, eagerness. To further fine-tune those answers, you can even use this Palette to get *percentages!*

Since you will probably be using your Compat Palette very often, have it copied on heavy paper. (Vellum Bristol card stock will run in most copy machines. Always show the *copyright permission statement:* You have permission to make *one* copy of each of the four Powell Meters: the Basic MultiMeter, the Time Machine, the Minute Multi-Meter and the Compatibility Palette.) Remember to laminate them.

Quick-Check For General "Compat"

Your first step is to become more consciously, objectively AWARE of the person/place/thing you are asking about. How does he/she/it fit into YOUR needs, your personality, your life-style?

Discover Your Perfect Mate

Check, as always, to see if this is the right time. With pendulum in your hand, ask: "Are these the right conditions to run a compatibility check on (say the name of person or thing)?" If the answer is YES, proceed; if NO, do as you did on page 32. Let's assume the answer is YES.

Now you're ready to check for *General Compatibility*. Place on the table the Compat Palette and a pad of paper. Touch your pendulum to the large darker center circle. Concentrate. You are dialing in the Compatibility Mode.

Ask, "Is (name of person or thing) generally compatible with me (in work, love, business, career, living arrangements, or whatever you are needing to know about)? (wait for answer) Assuming it was a healthy YES, now get the *Percentage* of Compatibility.

Percentage Of "Compat"

Touch your pendule to the Percent Mode, located in the "thumb hole" of your Palette. Now hold your pendule over the center *cross-lines* -⊕- (the *Starting Point*) and let the pendule swing. Note what number the pendule moves toward on the Percentage Arc. If between 70 and 80, you have 85% compatibility. Terrific! Write the person's name on your pad, along with that percentage number.

But if you received a NO answer, that you are not *generally compatible* with the subject (person, place, or thing), throw that name out! Out of your life. Even if that person is gorgeous, the job pays great bucks, the dog is adorable (he bites)!

Now you have the idea. But before you actually pendule for the perfect whatever, be sure to *read the entire chapter* and you'll see just how specific and in-depth you can go.

Nine Buttons For A Perfect Human Fit

Notice on your Compat Palette the BUTTONS for basic *categories*: Personal, Physical, Mental, Emotional, Social, Family, Spiritual/Beliefs, Career/Work, Financial.

For a more *detailed compat* rating, starting clockwise, touch your pendule to the first button—*Personal*. Dial in that category and think how it fits your particular situation with that specific person, place, or thing. Now, touch your pendulum to the Percent Mode. Next, hold it over the center cross-lines. Think. And let it swing. Write on your pad which Category and the percentage you received. Continue around the circuit of the Nine Buttons.

For each of the Categories, we'll give you many SubCategories to stimulate your own thinking, probing. This will help you to shed more light, to "pierce the persona." Read them through, and you'll see there is method in our madness. For almost always there is something you've overlooked and/or don't know about yet. We ask you to think about each of them, weigh carefully which characteristics are important to you; then pendule for percentages.

PERSONAL

Personality: Outgoing and a bubbly turn-on, inturning and a stony turnoff, outdoing one minute and inturning the next (drives you CRAZY!). A daytime person or a nighttime person. Life of the party or hides in the corner. Type A or Type B (two Type A's may be too much).

Habits: Slob or neatnik, too late or too early, couch potato or on the go-go (aerobics or just lies there!). Gourmet or gourmand, veggie or heavy meat-eater, smoker or non! (Says he/she's going to quit? Let 'em quit first, then you'll talk).

Communication: While one talks the other closes off, neither talk—both close off with a lot of negative inner dialogue going on. Both talk or scream or yell and nobody listens, both LOVE to talk and listen and BE. Both communicate subliminally without talk.

Power: Who's got to have it and who does not? Who has to have the last word? Who holds onto the TV remote control! Or who isn't assertive enough?

Honesty: They either have it or they don't! When they don't—run!!

PHYSICAL

While one likes the cold, the other likes the heat.

Health: One takes care of the body temple, the other doesn't give a good doggone—this goes for food, drink, drugs, sleep, exercise, you name it.

Sexual: Too much or too little. Too fast or too slow. Daytime, nighttime, or on the kitchen table!

MENTAL

Intellectual, street-smart or "not all there."

Ethical or cunning, forthright or devious; fair-minded or manipulating, reasonable or unreasonable; arrogant or humble or nicely-in-between.

Self-confident or self-abasing (with low self-esteem that needs constantly to be propped up—to where yours is sucked away! BEWARE of CO-DEPENDENCY in any relationship.)

Education: (what level and what kind): Accounting or basket weaving, mechanics or flute playing, engineering or epistemology, Ph.D. or dropout.

EMOTIONAL

Huggy bear or hands-off, cold as a cucumber or warm as an island breeze or smoldering hot.

Responsive to you, to other people, other races, other ethnic groups; to the heartaches of the world, the trees in the forest, the stars up above. Or tuned-out.

SOCIAL

Background: From across the railroad tracks or up on New York's Fifth Avenue.

Interests: Reading or TVing, theater-going or dancing or sitting-out, mountain-hiking or mall-strolling. Or sports sports sports versus bridge bridge bridge.

Sports: Viewing or doing, team or individual, wrestling or ice dancing. Inboard racing boat or 35-foot sloop or paddle and canoe. Scuba-diving or bobbing in the backyard pool.

Travel: While one loves to, the other is a stay-at-home; one wants to see all of Europe and the other wants to go up the Amazon, one loves trains and the other loves planes. One wants to drive there and see absolutely everything and the other one just wants to get there.

Vacation: Doing two weeks down the Colorado River or two weeks in the backyard hammock. Doing the hot spots in Paris or winter camping in the snows of Vermont.

FAMILY

Does he\she love your children, your pets, your friends, your family?

Children: He's got, she's got, neither have and both want; neither want—no room for kids in the "busy" life. This can be a real cruncher. I've seen divorces because one wanted and the other didn't—or one LIED before the wedding about "sure, love to have kids."

Adopting Children: Should you adopt or go for fertility procedures? If adopting, will it be a baby or an older child? Consider the woeful possibility of getting a baby addicted to drugs or alcohol.

Values: Can this person balance career and family? (Is this a *balanced* person?) If two people don't SHARE the same values, there is just no point in trying. It can only end badly, sadly. Too late.

SPIRITUAL \ BELIEFS

Soul Mate? Destined? Has this person come here to teach you something or you to teach him/her?

Religion: High Church and Ritual or "ecological spirituality" or not give a hoot.

Beliefs: One is Pro-Choice and free choice and the other is No Choice.

Fidelity: I only have eyes for you, or I love the one I'm near.

WORK | CAREER

Goals: To be the best bowler, the best jazz pianist, to win an Oscar, go for Olympic Gold, be the CEO, or no professional aspiration.

Work: His/Her line of work—fire fighter, test pilot, sexy model, politician, traveling sales rep, or 16-hour-a-day workaholic—can you happily, contentedly live with that? Perhaps you would work well *together*. Or "the other" doesn't work at all and you wish to golly he/she would!

Work Ethics: You're "promised" a pie-in-the-sky title, a raise, more authority and responsibility. Or—shoe on the other foot—YOU aren't giving full value... late, leave early, unfinished and unprofessional work.

FINANCIAL

Finances: Wanna make a million, got a million, haven't got a *sou*, or doesn't care about money. Is generous to a fault or has the first dollar ever; spends money like water or pinches to hear the buffalo squeal.

Investments: One wants the money "safely" in the bank and the other wants a high-return, high-risk mutual fund.

Property: One has and the other hasn't, neither have but both want, one wants but the other doesn't. And where you settle down can be a real divider.

Living: Your house or mine? Your kids or mine? Your town or mine? Decisions, decisions, decisions—let's call the whole thing off.

Gathering More Data

While you're checking for business remember to pendule about initiative, self-starting abilities; idea-person, sense of timing and priorities... and any other characteristics that you rank high in an employee/employer, in a partner/investor. Also check if this person works best in a group or alone.

For either a work or loving situation, is that person respectful of another's dignity? Or is condescending, criticizing? Is this a person who praises and trusts, allows you to think on your own, do your best work, be your best self?

And test for each of the following qualities, then BEWARE of their opposites: Loyalty, Responsibility, Dependability, Honesty. (You know where that person is "at." No tricks. No wondering, no going CRAZY.) Test to see if that person is: Considerate, Kind, Giving, Empathic, Sympathetic, In Tune with you!

Now you're getting more ideas for expanding your questions high and wide and deep.

> *Example:* When you touch your pendulum to the *Physical Button*, do an Energy Check. Ask, "Is (name)'s energy compatible with mine?" Answer NO.
>
> "Is (name)'s energy too high for me?" Answer YES.
>
> "Can (name)'s high energy be channeled productively into (my business, remodeling the house, exercising the dogs, cleaning out the garage)?" Answer YES.

Of course you might also ask if this high energy will keep you up at night? Disturb the quiet functioning of the office? Make you nervous, etcetera, etcetera! In other words, test before you leap over your head and are drowning—in love or in work.

Discover Your Perfect Mate

Better to be sure than sorry in a divorce court/bankruptcy court/or dog pound! (Remember that VCR? Check your new car purchase... before.)

It's All In The Percentages

Keep asking for percentages. It is up to you to decide just what level of INcompatibility you can live with. Be realistic in either direction. Rosy-colored love can't change a mismatch. Nor, on the other hand, should you dismiss a possible loving friend or good potential employee because one or two Compat Buttons show a low percentage.

Everything depends on just WHAT KIND OF RELATIONSHIP you are testing for. That low reading may or may not be significant.

Let's say the subject is an applicant for a job, who got a low percentage in Spiritual compatibility, but tests high "compat" in Mental and Personal. Would you hire that person? Of course! On the other hand, if you are looking for the perfect mate, you should most certainly look for one with a HIGH Spiritual Compatibility!

Fill In The Blanks

You're wondering about those two Blank Buttons? You think we left something absolutely fascinating out? Well, they're for you to write in special qualities dear to your heart. How about SENSE OF HUMOR? If you can't laugh together, forget it. Life becomes too heavy. Even in business, work becomes too dull, oppressive. How about SPIRIT OF ADVENTURE? Or POLITICS!! That could be a real blowup.

Don't forget to check where you like to live—city versus country versus "burbs." East Coast, West Coast or Dead Center. All these differences can break a relationship. *Chemistry is not enough. But it is a starting point—even in business. A starting point, not a decisive point.*

So don't get involved in love or live with someone or some thing if there isn't a HIGH COMPATIBILITY QUOTIENT. Don't settle for second best. Having settled, you may miss out on the "right" person just around the corner, just outside the office door, just down the hall from you.

Use Your Common Sense

Don't expect 100% compatibility unless you are a romantic dreamer, from another planet, or were cloned! A *few* differences in minor categories are fine for intrigue and stimulus, growth and fun! Yes, *opposites attract, but they don't fuse very well.*

After the excitement peters out, the coldness of INcompatibility sets in. There you are—caught. Don't let that happen. PENDULE FIRST.

It might be an illuminating exercise and certainly FAIR, to imagine yourself THE OTHER PERSON! How would YOU do if checked out? Remember, for COMPATIBILITY in anything both have to give and get!

In the end, it all comes down to a judgment call. You could give him\her a chance. With eyes wide open. Just be sure you don't sacrifice peace of mind or self-image.

May you live long, peacefully, and prosperously.

Taming The Wild Pendulum

Traveling In Your Time Machine!

Up to this point all of our questioning for the pendule has dealt with present time. But now you're ready to take off! We have turned your MultiMeter into a Powell Time Machine so you can peek into the past or zoom into the future.

Your Powell Time Machine is an eternal calendar you can use to move backward or forward in time (see Diagram 18). You may go into the past to discover when something took place, or into the future to discover when something *will* take place.

As an example, perhaps in Chapter 6 you found that your test subject had a broken bone. It might

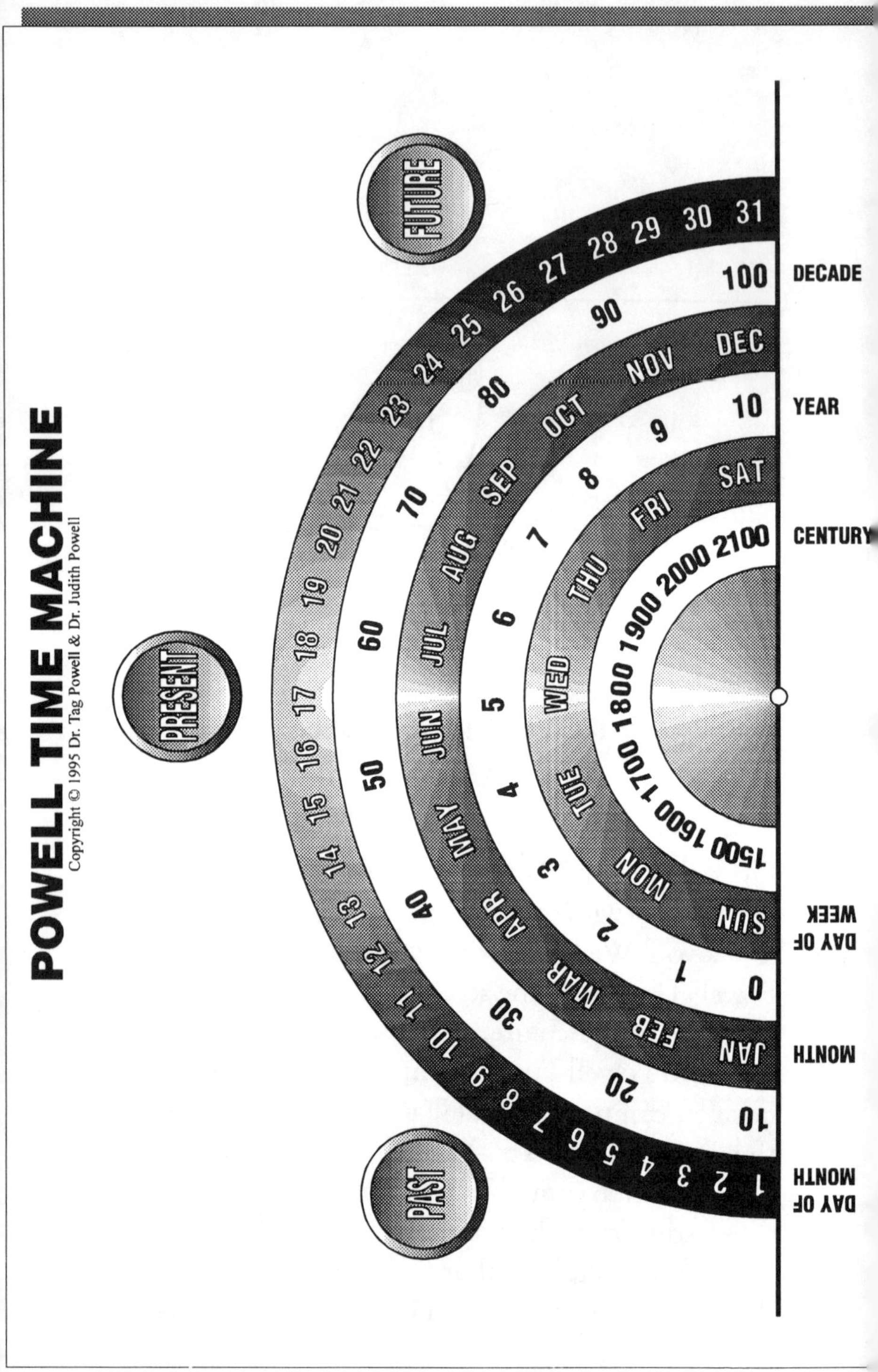
POWELL TIME MACHINE
Copyright © 1995 Dr. Tag Powell & Dr. Judith Powell
FUTURE
PRESENT
PAST
DECADE
YEAR
CENTURY
DAY OF WEEK
MONTH
DAY OF MONTH
1 2 3 4 5 6 7 8 9 10 11 12 13 14 15 16 17 18 19 20 21 22 23 24 25 26 27 28 29 30 31
10 20 30 40 50 60 70 80 90 100
JAN FEB MAR APR MAY JUN JUL AUG SEP OCT NOV DEC
0 1 2 3 4 5 6 7 8 9 10
SUN MON TUE WED THU FRI SAT
1500 1600 1700 1800 1900 2000 2100

DIAGRAM 18

Traveling In Your Time Machine!

be fun to move back in time to find just when this bone was broken. You may also want to discover when your allergies started, when a negative limiting belief began, or perhaps when you've lived before.

The pendule will also be your vehicle to take you where you want to go into the future. Go where the answers lie. That is, which day of the month, the actual month, year (decade and century) when you should move into a new career, to a new home, to a new country, even to a new planet!

Looking Into The Future

As if looking into a crystal ball, you can ask your pendulum for answers predicting your future; the future of a state, country, planet, species!

For example, ask your pendulum the question, "Should I look for another career?" If your pendulum swings YES, then you may choose to hop aboard the Powell Time Machine.

With your pendule hanging above the Powell Time Machine, touch it to the Future Button. Then touch it to the Day-of-Month Band, then hold it over the *Starting Point* (small center circle), and ask, "Which day of the month is best for me to make my move to a new job?" (wait for answer)

After the pendule starts to swing, you can use a straight edge or ruler to guide you in accurately reading the number it is swinging toward.

Then touch your pendule to the Month Band and ask, "Which month is best when I should make my move?" (wait for answer)

Then touch your pendule to the Year Band and ask, "Which is the best year when I should make my move?" (wait for answer)

If you live in an environmentally dangerous area, you may ride the Time Machine and ask the pendulum to predict when the volcano will erupt; the hurricane will hit; the earthquake will shatter; the fire, mudslide, or flood will consume whatever is in its path.

Predicting The Elections

Our own research has shown the Powell Time Machine to be more accurate about the past than the future. Yet, we and our students have had several impressive successes with future pendulum predictions.

I (Judi) even predicted the outcome of the 1993 U.S. Presidential election *before* the political tides had turned. I predicted Bill Clinton would win the election when Bush was the favored to win by a land slide.

We were in Cyprus conducting our "Mind Mastery" and "Taming the Wild Pendulum" semi-

Traveling In Your Time Machine!

nars. Our hotel did not get any English-speaking channels, nor did they get satellite for CNN. We were cut off from world news. Over dinner one night, our friend, Phivos, asked whom we thought would win the U.S. election. It became clear to us that the outside world is very interested in who gets in "our" White House, since that person affects foreign policy.

Well, right away Tag said, "Bush is a shoe-in." This was common thought in the polls when we left for our travels June 1st. Phivos said, "Why not ask your pendule?" So, I tore three identical pieces out of my notepad and wrote one candidate's name on each: Bush, Perot, Clinton. I turned them over, mixed them up and laid the papers next to each other on the table.

I then began the predicting process. With pendule over each piece of paper, I asked,

> *Paper 1:* "Is this person going to be the next President of the United States?" YES at first, then it swung NO.
>
> *Paper 2:* "Is this person going to be the next President of the United States?" YES!
>
> *Paper 3:* "Is this person going to be the next President of the United States?" NO.

Then I turned over the papers to find out which name was written on which paper: paper #1 Perot;

paper #2 Clinton; paper #3 Bush. What a shock!!! Clinton to win the White House? How could that be. Tag felt the pendulum was wrong, so we asked it again and again and again... until the pendule "came up" with the answers *we thought* were correct!

The moral of the story is to trust your pendulum responses, for it is only a means of getting to the REAL truth.

One explanation for the vagueness or wishy-washiness about predictions can be that the future is not fixed, as some believe, but is ever-changing. The very decisions and actions you take at this moment influence future moments. Another reason can be the scientific theory that we live in parallel universes—living simultaneous lives—where each of your various decisions and outcomes can be played out.

Help us test out a theory we're working on to make predicting the future more accurate. Here's how: *when you ask a futuristic question, qualify it with "for this lifetime," or "for this parallel universe."* Write us your results so we may add it to our data.

Discovering The Past

Unlike our future, at least our past is fixed... unless you accept the possibility of true time travel which can open a vast paradox!

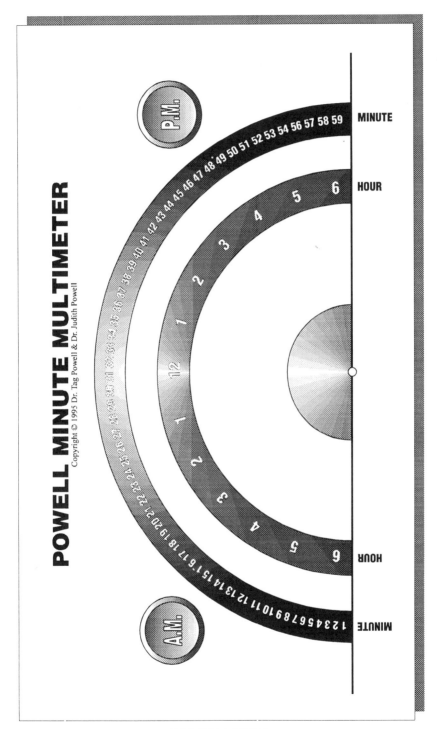

DIAGRAM 19

Your Present Past
Discovering The Beginning Of A Problem

There is in our memories a recorded history of everything that has happened to us in this lifetime. Tapping into these mental data banks can be easily accomplished with your pendulum. One of the factors which explains the amazing accuracy of recalling past experiences of yourself and others is the emotional factors. Strong emotions create major impacts on the brain neurons, almost like file tabs in life's filing cabinets.

Take any major health problems, the strong emotional impact of being sick or hurt leaves a lasting imprint, or "groove" if you will, located in the brain. For this reason, we can pendulum-dowse the past.

Let's take the broken bone we mentioned earlier, as our example. The break probably caused great pain—a strong impression in the memory. So, when we are using the pendule to scan the body matrix of another person, the moment of the "hurt" is quite easy to find.

When doing a health study, having located a broken bone on the skeletal structure of your subject (using Diagram 10), you can find when the break took place using the Powell Time Machine.

Traveling In Your Time Machine!

With your pendulum hanging above the Powell Time Machine, touch it to the Past Button. Then touch it to the Decade Band, and hold it over the Starting Point (small center circle). Ask, "What decade did this bone break take place?" (wait for answer) Note your answer on a piece of paper.

Then touch your pendule to the Year Band, and ask, "Which year did this bone break occur?" (wait for answer) After the pendule starts to swing, you can use a straight edge to guide you in accurately reading the number it is swinging toward. Write your answer on the piece of paper.

Before you continue, after you have found the year, you can double-check your accuracy by going back to Diagram 10 of the human skeleton.

Ask, "Is there a break in the leg in (date)?" (wait for answer) Always use a date one year earlier than the number "divined" with your pendule. By going one year earlier, 1962 for our example, you should get a NO reading.

You can further discover the month and even the day depending on your desire for pinpointing the event. For example, the broken bone took place in 1963... but when in that year?

Next touch your pendule to the Month Band, and ask, "Which month did the bone break take place?" (wait for answer) Note your answer on the paper.

Lastly, touch your pendule to the Day of the Month Band, and ask, "Which day did the bone break take place?" (wait for answer) Use your straight edge to guide you to accuracy. Note the answer on your paper.

You can even get into the nitty-gritty and divine for the actual hour, minute, and second the bone break took place. (see Diagram 19)

As you can see, using the Powell Time Travel Machine in the health arena has almost endless uses.

Using this procedure you can also discover when life-changing events (positive and negative) took place. By finding the event in the past perhaps it will give you an understanding of how you can change your reaction in the present—which can ultimately affect your future.

Questions to help you may find a date locating a pivotal event in your thinking:

When did (I or subject) start having this fear... distrust... belief... talent... lack of talent... negative attitude... positive attitude... feeling about money? (see our books *Think Wealth* and *The Science of Getting Rich*).

Many psychologist believe that by discovering the trauma of a negative event of the past it will relieve the problem in the present.

Discovering Your Past Lives

The belief of having lived before this lifetime is a fascinating spiritual and personal study

We have done thousands of past life regressions, both for groups and one-on-one sessions. It is a fact that most people have easily recalled memories of previous times when they have lived before. Debate of whether these memories are actual events or figments of our imaginations will probably rage on forever. We may even go back to a past life where we had this same discussion!

Doing a past life regression is too involved to go into in this book. For a training and script on how to do a past life regression, read our book *ESP For Kids*—don't let the title fool you—it's for adults too!

What Is The Value In Past Life Regression?

Many individuals are being helped by therapists/hypnotists/psychologists/psychiatrists who have help alleviate fear, guilt, and doubt through having the person recall a past life. Whether the past life is real or imagined is not as important as the fact that this form of therapy is effective for a large number of people.

As we may have had many different past lives, discovering which one has the cause—the *original* specific problem—can be very time-consuming. Using the Time Machine, you can zero in on the exact lifetime containing the *root cause*.

Here are some sample questions to ask yourself using the *full* Time Machine procedure as in the broken bone scenario. Except this time, begin with the Century Band, then chunk down from there.

- ✧ "In a past life, which century did I create the problem?"
- ✧ "Which year should I go to make the greatest changes in my attitudes?"
- ✧ "Which year should I go to recall and release my latent talents?"
    "... to discover my true destiny?"
- ✧ "Which year should I go to heal the cause to my health challenge?"

Once you discover a past life, go back to that time using a guided regression meditation (see *ESP For Kids* for script), or use your dreams—to release the healing knowledge to your *present* outer- and inner- conscious levels.

Enjoy. We wish you good timing in discovering your past, thereby accepting your present, thus creating a better future!

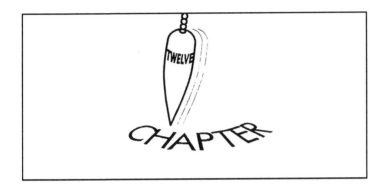

The Magic Doorway: Opening Your Chakras

The Chakras are called the seven basic energy sensors of the body. The oft-mentioned *Kundalini experience* is when all the body sensors are open and in perfect balance. It is said that when the Chakra Centers are open and balanced, we are at peace and harmony with the world around us. If you or any of your loved ones are experiencing an out-of-balance feeling, seeming to be agitated or out of step with life, the answer may be to open your chakras.

If the concept of chakras seems too foreign to you, look upon it as a metaphor for rebalancing the *body's energy*. The Chinese call this energy *Chi*.

By knowing about and developing your chakras (whether you consider them actual or metaphoric), you can become not only a more harmonious being, but a more successful, creative, productive one. With the energy sensors in balance, from the root chakra to the crown chakra, you will experience greater insight and increased wisdom.

Since many books have been written on the subject (some are listed in the Bibliography), we are going to cover here only the *seven Basic Chakra Centers* (see Diagram 20), and give what we hope is a simple and usable meaning for each. You may then use your pendulum to *test* for sluggishness, then open or rebalance your chakras if necessary.

So just what are *chakras* (Sanskrit to mean *wheel-like whirling vortices*). The chakras are the seven centers of Kundalini, of "spiritual energy" in the body. These vortices are the gateways which allow in the flow of life-giving *prana/energy*. That is, energy which could be considered LIFE itself. They are the *etheric double* of man, the invisible part of the physical body. They are our bridge to the beyond.

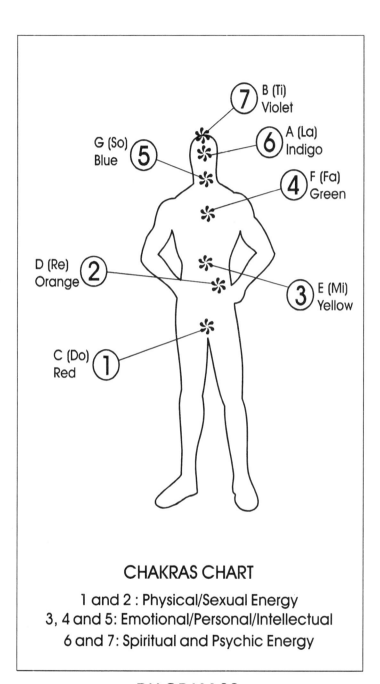

It is believed that within the chakras lie connecting points of energy centers for all three bodies of our being—the physical; the emotional (personal, mental, and intellectual); and the spiritual.

Clairvoyants able to see "auras" report that the chakras are pastel-colored *swirling whirlpools of light*, perpetually rotating in a *counterclockwise* fashion. The colors are of the rainbow, expressing the spiritual light from the Primary Source.

In the individual who is physically weak or spiritually underdeveloped (or emotionally depressed) the seven centers are said to be small, dull, sluggish vortices that are moving slowly, out-of-sync. The aura is seen to have shrunk down closer to the body, the colors darker, often muddied. The gateways are allowing just enough energy in to sustain life.

Yet when the centers are awakened and vivified, the individual becomes physically energized, emotionally responsive, and spiritually enlightened. The chakras are seen then as large, glowing pulsations of light. The aura radiates out, bathing everyone around in a happy, serene glow.

It is said that the aura can change from day to day, depending on the *ephemeral* health of that person. Some who cannot "see" the aura, "feel" the vibrations (popularly called "vibes"). Everything has its vibe: a house, a tree, a car.

The Magic Doorway: Opening Your Chakras

Each chakra displays its own predominant color energy (see Chakra Chart) and musical note (spiritual level); its own personal and intellectual tone (emotional level); with each being linked to one particular organ or glandular function (physical level). All the chakras are interconnected; as all levels of our being are interconnected; just as all beings are interconnected.

The Interconnection Of Physical, Mental, and Spiritual

The *energy centers* vary in size and in brightness in different people. Even in the same person some of the centers may be more developed than the others. If an individual excels in a particular quality, that center will be much enlarged and more radiant than the others.

Chakras 1 and 2 are primarily concerned with Physical Energy: sexual energy (serpent-fire) from the Earth; and vitality from the Sun.

Chakras 3, 4, and 5 are primarily connected to Emotional Energy: the Personal and Intellectual forces.

Chakras 6 and 7 are developed only when a certain amount of Spiritual Energy is involved. They are more complex in nature, encompassing multiple avenues for growth from the Spiritual Plane.

Testing Your Energy Levels In 3 Steps

Now let's pick up your pendule and testing those chakras!

STEP ONE: Testing for Sluggish Chakras.

To discover which chakras may be in need of "energy repair," use as your *map* the Chakra Chart (see Diagram 20). Make a photocopy of this chart and date it, so you can keep track of your progress in opening up your chakras. Make several copies, so you'll have one for each time you test your energy level. As usual, you'll also need a pointer and highlighter pen.

The *serpent-energy*, *Kundalini*, rises from the base of the root chakra upward to the crown chakra. So it is this upward flow that becomes the order in which you will test for sluggish chakras.

Touch your pointer to the *root chakra* on the chart and, holding your pendule over the

The Magic Doorway: Opening Your Chakras

Basic MultiMeter, ask, "Is this chakra in need of energy repair?" (wait for answer and mark it on the photocopy)

If the pendule swings YES, touch your pointer to the % Percentage Button on the Basic MultiMeter© and mentally dial in that Mode. Now with your pendule held over the Starting Point, ask, "What percentage of blockage has this root chakra?" (wait to see what number your pendule swings toward, mark it on the photocopy) Say it reads 4—this would mean 40% blockage.

Continue on up the Chakra Chart, touching each chakra, marking your answers on the photocopy.

STEP TWO: Analyzing the Percentages on All Levels—Physical, Emotional, Spiritual.

When you pendulum-dowse and find certain chakras give you an answer of "closed 50%" or even a higher percentage, these chakras are in definite need of opening. This could indicate corrections are needed on several levels.

STEP THREE: Analyzing the Problem.

Read the following descriptions on the seven chakras and their three levels of influence. Then take appropriate steps to bring back into balance the chakras which penduled more than 20% sluggishness.

Analyze the percentage of repair needed, then correct the problem or challenge through: 1) proper medical or herbal treatment; 2) correct thinking; 3) proper behavior; and/or 4) mental visualization using color energy.

1. Root Chakra:

✧ *Physically,* your nerve energy disseminating from the base of your spine may be blocked, out of alignment; or you may have some health problems in your sex organs.

✧ *Emotionally,* you may be thinking or expressing sexual lust or guilt, or possibly stuck in the overactive "hormonal" action of a teenager.

✧ *Spiritually,* you may not be working out of the "love" mode when reacting to life, love-making, or even eating (gobbling food, rather than enjoying/appreciating it!) Think loftier thoughts. Aspire!

2. Spleen Chakra:

✧ *Physically*, your spleen may be over- or under-producing lymphocytes, monocytes and plasma cells; or perhaps it's not properly filtering your blood for bacteria, debris, and parasites.

✧ *Emotionally*, you may be expressing too much anger ("venting your spleen!"). You may be too often "giving hell" verbally to your family, friends, neighbors; or, just the opposite, you may be holding too many emotions inside. Spiteful or antisocial behavior, held-in anger or umbrage, can block this chakra.

✧ *Spiritually*, you may not be filtering out debris from your life; like old thought patterns, and replacing them with mind-expanding beliefs.

3. Navel (Solar Plexus) Chakra:

✧ Physically, the autonomic nervous system is primarily concerned with keeping the body systems running smoothly, day in and day out. It is comprised of the sympathetic and parasympathetic nervous systems, which may be "inflamed." This could be from excessive, running-rampant emotions. The solar plexus is like the *emotional brain* of the body.

✧ *Emotionally*, your brain and mind may be dulled, under- or overstimulated. You may be running on logic alone.

✧ *Spiritually*, you may be leading your life using the act-then-react method; using only your intellect (brain), rather than incorporating your intuition (mind).

4. Heart (Thymus Gland) Chakra:

✧ *Physically*, your thymus gland may be over- or under-producing white blood cells and T-cells, which keep your body safe from disease. Your immune system may be in need of energizing.

✧ *Emotionally*, you may be insecure about your situation; love, money, the future.

✧ *Spiritually*, you may not be connected to your Higher Source, therefore you are cut off from the security of "Pure Love."

5. Throat (Thyroid) Chakra:

✧ *Physically*, your thyroid gland may be over- or under-producing the hormone *thyrozine*, which sets your basal metabolism rate.

✧ *Emotionally*, if overproducing, you may be too excitable, sometimes "over the edge." If underproducing, you may tend to be sluggish, no energy, feeling and acting dull. You may not be able to bring your concepts, desires and dreams to fruition—you may be failing to accomplish your goals.

✧ *Spiritually*, you may not have connected with your destiny, your true purpose this time around. In short, you may have missed the boat if underproducing, and shot right over it into the shark's jaw if hyperproducing!

6. Brow (Pineal) Chakra:

✧ *Physically*, your pineal gland may be over- or under-actively producing the hormone *melatonin*, which regulates the ovaries; and may help to regulate the pituitary gland. The pineal gland responds to environmental lighting, and is believed to be the seat of the body's *biological clock*.

✧ *Emotionally,* you may be thinking that time is "running out" for you to fulfill your goals, your purpose in life; your destiny.

✧ *Spiritually,* you may not be paying attention to (seeing with the "inner light") those whisperings of intuition giving you direction. It is this chakra that is sometimes called *"the third eye."* Through it mystics are said to envision marvelous sights.

7. Crown (Pituitary) Chakra:

✧ *Physically,* your pituitary gland may be over- or under-actively producing the Growth Hormone, *somatomedin.* More accurately, it's the *hypothalamus gland* that is involved with this chakra. It is the Master Control Center for all the other glands. It gives feedback to them to increase or decrease hormonal production. This occurs especially when the body is under stress. Its full functioning is still a scientific mystery.

✧ *Emotionally,* you may not be "growing"—controlling and expressing your creative, imaginative side. You may not be the *master* of your own life; not being all you can be. Excel!

The Magic Doorway: Opening Your Chakras

✧ *Spiritually*, you may not be meditating—getting *feedback* from your Master self—on how you can create a better world in which all of us can live. Create a legacy! (and we don't mean a baby!)

STEP THREE: Opening and Balancing the Seven Chakras.

To balance the three levels of each chakra:

✧ PHYSICAL: We balance the *Physical chakras* by what we put into our bodies—the food and water we take in; the quality and quantity of the air we breath; and the quality and quantity of rejuvenating sleep we get. We must gain control over the lust for sustenance, sleep, sex!

✧ *EMOTIONAL:* We balance the *Emotional chakras* with what we put into our hearts and minds and the emotions, desires created from this. By the daily conversations we engage in or listen to (TV and talk-radio included); the thoughts we think, the inner self-talk with which we bombard both our inner and outer consciousness. The quality of the music we listen to; the books we read; and quality of the music we listen to; the books we read; and the lectures or seminars we attend. Let us gain control over our emotions and intellect!

✧ *SPIRITUAL:* We balance the *Spiritual chakras* with what we stir the fires of our souls; the control we exercise over our base natures; the quiet or meditative time we allow ourselves to connect our being with the Source; the flashes of insight we observe; the wisdom we internalize into our very core.

Visualize perfection and you will become it! We both wish you a balanced life.

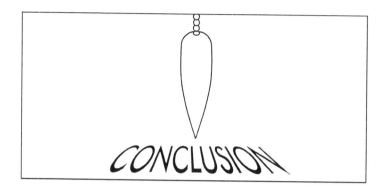

CONCLUSION

Where Do You Swing From Here?

So you have the pendulum swinging and the bobber bobbing and would now like to continue your study? You want to keep using, exercising your inner-conscious mind to aid in your problem solving—in life, love, and business?

One of your best bets would be to join The American Society of Dowsers (A.S.D.) in Danville, Vermont. The cost is minimal, under thirty dollars per year. Membership is open to professional and amateur dowsers alike, as well as anyone interested in the art of dowsing. The society has local chapters in just about every state and in some cases, several per state. There are also dowsing societies in almost all westernized countries.

You can attend local monthly meetings, as well as any of the several dowsing conferences held yearly around the United States. Major dowsers get together each year at the Annual Convention of the American Society of Dowsers at their headquarters in Danville.

If you don't travel, the A.S.D. will keep you informed through *The American Dowser Quarterly Journal* plus a bimonthly newsletter *The Dowser's Network* (quarterly). Both the journal and newsletter cover dowsing techniques for the beginner and the veteran; including latest newspaper reports on dowsing, as well as the latest scientific research.

Each issue contains articles from dowsers around the globe. Photos and illustrations add to your enjoyment and greater understanding of this art of the dowser.

Happy Penduling!

Recommended Reading

Recommended books to read and audiocassettes to listen to for Balancing the Chakras.

To Balance The Physical Chakras:

✧ The *Science of Well-Being* book by Wallace D. Wattles and Dr. Judith Powell $8.95;

✧ *Physical Well-Being audiocassette* (meditation & subliminals) by Dr. Judith Powell $9.95;

✧ *Color Healing audiocassette* (guided visualization & subliminals) by Dr. Judith Powell $9.95;

To Balance The Emotional Chakras:

✧ *Silva Mind Mastery for the '90s, book* by Drs. Tag and Judith Powell $17.95;

✧ *Emotional Self-Control audiocassette* (meditation & subliminals) by Dr. Judith Powell $9.95;

✧ *Mental Power audiocassette* (meditation & subliminal) by Dr. Judith Powell $9.95;

✧ *Personal Growth audiocassette* (meditation & subliminals) by Dr. Judith Powell $9.95.

To Balance The Spiritual Chakras:

✧ *The Science of Becoming Excellent book* by Wallace D. Wattles and Dr. Judith Powell $8.95;

✧ *Increase Your Spiritual Awareness audiocassette* (meditation & subliminals) by Dr. Judith Powell $9.95.

To Balance All The Chakras:

✧ *Up Your Energy* award-winning audiocassette (guided visualization & subliminals) by Dr. Judith Powell $9.95.

General Enlightenment

✧ *Think Rich... Put Your Money Where Your Mind Is!* by Dr. Tag Powell $8.95;

✧ *ESP for Kids* by Dr. Tag Powell and Carol Howell Mills $12.95;

✧ *The Science of Getting Rich* by Wallace D. Wattles and Dr. Judith Powell $8.95;

✧ *How to Discover Your Soul Mate* audiocassette and booklet by Dr. Judith Powell $12.95;

Order from the publisher:
Top Of The Mountain Publishing
P.O. Box 2244
Pinellas Park, Florida 34664-2244 U.S.A.
Fax (813) 391-4598
Phone (813) 391-3843

(Add $4.00 for first item; $1.00 for each additional item. Your order will be sent to you by *U.S. Priority Mail* in the continental United States. For Foreign orders, please write or fax the publisher for shipping prices.)

Write, Phone, or Fax for FREE Catalog!

Resources

American Society of Dowsers, Inc.
P.O. Box 24
Danville, Vermont 05828 U.S.A.
Fax: (802) 684-2565
Phone: (802) 684-3417

Johnson Smith Company
P.O. Box 25500
Bradenton, Florida 34206-5500 U.S.A.
Fax: (813) 746-7962
Phone: (813) 747-2356

Radon Gas Detection Kit
P.O. Box 562
Jonestown, Pennsylvania 17038 U.S.A.
(Must order through the mail)

Top Of The Mountain Publishing
P.O. Box 2244
Pinellas Park, Florida 34664-2244 USA
Fax: (813) 536-4598
Phone: (813) 391-3843
Write, Phone, or Fax for FREE Catalog.

About the Authors

Dr. Tag Powell

Dr. Tag and Dr. Judith Powell are living examples of what they preach: that by learning to scientifically train and focus more of your mind, you can become a happier, wealthier, and more successful person. Their seminars have been applauded from New York to Hong Kong to Greece.

Dr. Judith Powell

As two leading human potential speakers and trainers in the United States, Tag and Judi have expertly guided thousands of people along this path of excellence for almost two decades. Their many self-help seminars range from learning to love yourself and increasing your financial status to state-of-the-art techniques of direct response marketing.

As lecturers, they both have received numerous awards and top honors. Tag was awarded the Silva Method's prestigious President's Cup as the outstanding lecturer in the United States. Judi has also won national top honors, including the Benjamin Franklin Award for best audiocassette, *Up Your Energy* at the American Booksellers Convention.

About The Authors

They are currently traveling worldwide presenting: Think Money, SALES POWER Program, Internet Marketing and Electronic Publishing, as well as the state-of-the-art Direct Response Marketing for Business Success.

As authors, the Powell's books have already been translated into over twenty-five languages. They both have authored and narrated bestselling life, business-improvement audiocassettes available in better bookstores around the globe. The Powells write numerous magazine articles including two monthly columns for international publications, and they continue to share their insights in new books and seminars.

Both Tag and Judi have earned their doctorates in Psychorientology (the highest certification in their field) from the Institute of Psychorientology in Texas. Tag studied with Stephanie and Dr. O. Carl Simonton, specialist in the field of helping cancer patients through the use of visualization.

He and Judi have earned their Master's Certification in Neuro Linguistic Programming (NLP), and are also NLP Associate Trainers, certified by the Society of Neuro Linguistic Programming through the Southern Institute of NLP. Tag and Judi are also Past-Life researchers and trainers for the Association for Past-Life Research and Therapies, Inc., and lecture at their conventions.

Hosts of their own television show, *It's All In Your Mind*, Winner of Cable's Spotlight Achievement Award for Best Educational Cable TV Show, they are also favorites of other top radio and TV talk shows around the world.

Judi was recently featured on the national NBC television show *The Other Side*. She explained how anyone can discover their perfect soul mate with mind power!

The Powells bring wit and humor to very important topics and zest to their seminars and lectures. At one point, Tag was a recognized New York actor and comedian. His diverse background includes: business management, President of an advertising agency, and he has even designed rockets.

Before meeting up with Tag, Judi earned her B.A. in Color Design and Business Management in Detroit, Michigan, and was then in the process of becoming a medical doctor. Love brought them together, and now Tag and Judi both help people to help themselves—physically, mentally, financially, and spiritually.

Tag and Judi are just as diversified today, both being authors, seminar leaders, and trainers.

Their recent speaking/training engagements include Los Angeles, Hartford, Chicago, Houston, Atlanta, and overseas in Indonesia, Singapore,

About The Authors

Bangkok, Frankfurt; at the University of Hong Kong, and the Malaysian Institute of Management in Kuala Lumpur. They have just returned from touring in Hungary, Cyprus, and Greece. Their seminar in Athens alone drew over 2,000 people.

When not traveling and lecturing around the world, Tag and Judi make their home in St. Petersburg, Florida, with their enlightened Scottish Terriers: Master, Buddha, and Isis.

POWELL SEMINAR INFORMATION
To schedule the Powells to conduct a seminar
for your group or organization
in the U.S.A. or around the world;
and
For a listing of seminars conducted
by the Powells,
or for a seminar date schedule:
Call, write, or fax
Top Of The Mountain Publishing
P.O. Box 2244
Pinellas Park, Florida 34664-2244
Fax: (813) 391-4598
Phone: (813) 391-3843

Bibliography

Askew, Stella. *How To Use A Pendulum.* Long Creek, South Carolina: Tri-State Press.

Cameron, Verne L. *The Dowser's Handbook Series No. 1, Map Dowsing.* Santa Barbara, California: El Cariso Publications, 1971.

Cameron, Verne L. *Aquavideo: Locating Underground Water.* Santa Barbara, California: El Cariso Publications of Life Understanding Foundation, 1978, revised edition.

Cox, Bill. *Techniques of Pendulum Dowsing.* Santa Monica, California: Forces, 1977.

Finch, Elizabeth and Bill. *The Pendulum & Your Health.* Cottonwood, Arizona: Esoteric Publications, 1977.

Finch, W. J. "Bill." *The Pendulum & Possession.* Cottonwood, Arizona: Esoteric Publications, 1975 revised edition.

Graves, Tom. *The Dowser's Workbook: Understand & Using The Power Of Dowsing.* New York, New York: Sterling Publishing, Inc., 1989.

Graves, Tom. *The Elements Of Pendulum Dowsing.* Longmead, Shaftesbury, Dorset, England: Element Books Limited, 1988.

Hoffmann, Wendell H. *Hoffman Students' Handbook: Pendulum and Projection.* Utah: Wendell H. Hoffmann, 1992.

Howells, Harvey. *Dowsing For Everyone: Adventures and Instruction in the Art of Modern Dowsing.* Lexington, Massachusetts: The Stephen Greene Press, 1979.

Howells, Harvey. (1982) *Dowsing: Mind Over Matter.* Lexington, Massachusetts: The Stephen Greene Press, 1982.

Jaegers, Beverly C. *The Extra Sensitive Pendulum.* Creve Coeur, Missouri: Aries Productions, Inc., 1972.

Leadbeater, C. W. *The Chakras.* Wheaton, Illinois: The Theosophical Publishing House, 1980, third edition.

Lübeck, Walter. Das Aura Heilbuck, Aitrang, Germany: Windpferd Verlagsgesellschaft mbH., 1992, third edition.

Lonegren, Sig. *The Pendulum Kit.* New York, New York: Simon & Schuster, Inc., 1990.

Nielsen, Greg. *Beyond Pendulum Power.* Reno, Nevada: Conscious Books, 1988.

Nielsen, Greg and Polansky, Joseph. *Pendulum Power.* Rochester, Vermont: Destiny Books, 1987.

Olson, Dale. *Advanced Pendulum Instruction & Applications.* Eugene, Oregon: Crystalline Publications, 1991.

Powell, Dr. Judith . *How to Discover Your Soul Mate.* Largo, Florida: Audiocassette and Booklet. Top of the Mountain Publishing, 1995, second edition.

Powell, Dr. Judith and Wallace D. Wattles. *The Science of Getting Rich.* Largo, Florida: Top of the Mountain Publishing, 1993, second edition.

Powell, Dr. Judith and Wallace D. Wattles. *The Science of Becoming Excellent.* Largo, Florida: Top of the Mountain Publishing, 1994, second edition.

Powell, Dr. Judith and Wallace D. Wattles. *The Science of Well Being.* Largo, Florida: Top of the Mountain Publishing, 1994, second edition.

Powell, Dr. Tag. THINK WEALTH: *Put Your Money Where Your Mind Is!* Largo, Florida: Top of the Mountain Publishing, 1991.

Powell, Dr. Tag and Carol Howell Mills. *ESP FOR KIDS: How to Develop Your Child's Psychic Abilities.* Largo, Florida: Top of the Mountain Publishing, 1993.

Powell, Dr. Tag and Dr. Judith Powell. *Silva Mind Mastery for the '90s.* Largo, Florida: Top of the Mountain Publishing, 1995, fifth edition.

Stine, G. Harry. *Mind Machines You Can Build.* Largo, Florida: Top of the Mountain Publishing, 1995, third edition.

Willey, Raymond C. *Modern Dowsing: The Dowser's Handbook.* Phoenix, Arizona: Esoteric Publications, 1976.

Woods, Walter. *Letter to Robin—A Mini Course in Pendulum Dowsing.* Walter Woods, 1993.

INDEX

A

allergies 31, 103
    other than to foods 109
    questioning, specifics 108
    testing for 103
        foods and liquids 104
        procedure 104
            school of thought #1 104
            school of thought #2 106
            school of thought #3 107
allergy tester 42
anatomy charts .... *See* health, graphics
ASD .... *See* The American Society of Dowsers
Aura Meter 114

B

Backster, Cleve 82
    plant and cell research 82
base number 75
Basic MultiMeter .... *See* MultiMeter, Basic
bobber (dobber) 111, 117, 121
    creating it 115
    definition 112
    graphic 113, 118
    training it 114
    usage 123
        lotto numbers 123
boustrophedon (ox-plow pattern)
    graphic 55

C

Cameron, Vern L 50, 114
Cartographic dowsing .... *See* map dowsing
Chakras
    definition 163
    graphic 165
    interconnection 167
    to balance 175
        recommended reading 179
    to open 170
        brow 173
        crown 174
        heart 172
        navel 171
        root 170
        spleen 171
        throat 173
    to test, 3 steps 168
Chi 164
common sense 148
compatibility 131
    analyzing 147
    chart graphic 135
    nine buttons 138
        emotional 141
        family 142
        financial 144
        mental 141
        personal 139
        physical 140
        social 142
        spiritual beliefs 143

compatibility, continued
    work/career 144
    quick-check 136
    percentages, figuring 137
conditions determining correctly 32
copyright, permission 136
counting procedure 74

D

diverter .... *See* noxious rays, how to eliminate
dobber .... *See* bobber
dowsing
    definition 16
    devices 16, 49
        forked stick (Y-rod) 16, 49, 111
        graphics 17
        L-rods 16, 49, 111
        plumbob 16
        wand 16, 112
    distant. .... *See* map dowsing
    map. .... *See* map dowsing

E

etheric double 164

F

facilitator
    as in psychotherapy 13
food samples. .... *See* allergies, testing for
forked stick 16, 49, 111
Foucault pendulum 15

G
goal setting 122

H
health
- graphics, how to use 90
  - body organs 97
  - digestive system 99
  - female 92
  - head and neck 96
  - heart 98
  - male 93
  - muscle 94
  - skeleton 95
- testing 84
  - how to 86
    - example, John's liver 88
  - proxy method 85, 106

health band, using 85
- graphic 86

health-check questions 90
health disclaimer 83, 89

I
illnesses
- detecting 81
  - the legality 83

illustrations, list 5
imprint .... *See* programming
inner conscious 11, 21, 28, 33, 42, 44, 80, 110
- intuitive (right brain) 12, 16, 23

inner conscious, continued
inner self. .... *See* inner conscious
intuition 12, 16, 23

J
Judi's solution 45

K
Kundalini 164

L
lamination of charts 69
left brain .... *See* outer conscious
legality of dowsing 83
L-rods 16, 49
Lotto numbers
   how to get
      bobbing 123
      penduling 126

M
Magic Pendulum 41
map dowsing 49
   accuracy level 51
   experiments, A.S.D. 52
   find lost treasures 52
      publisher's treasure hunt 59
         map graphic 60
   for missing person 50
   for noxious rays (radon gases) 56
   how to, four steps 53
      Colombia example 50

map dowsing, continued
    map size 51, 53
mate, discover 131
metaphysics 42
mind
    inner-conscious level .... *See* inner conscious
    outer-conscious level .... *See* outer conscious
MultiMeter
    Basic 67
        number counting .... *See* Thought Buttons
        percentages 77
        test mode 72
        thought buttons .... *See* Thought Buttons
        with Health Band 85
            graphic 86
        X 10 Button .... *See* Thought Buttons
        X 100 Button .... *See* Thought Buttons
    copyright 69
    design 68
    graphic 70, 86
    lamination 69

N

no, programming 29
noxious rays 55
    discover, how to 56
    diverting, how to
        two methods 58

Index

O

outer conscious (left brain) 12, 16, 28, 44
   analytical 12, 16
      trick, how to 45
ox-plow pattern .... *See* map dowsing, how to

P

past .... *See* Time Machine
past life 160
past life regression 161
   script 162
pendule body 19
pendulist 33, 43, 50
pendulum
   accuracy 44
   controlling force 42
   definition 15
   history 15
   operate, how to 26
      graphic 27
   homemade 20
      make, how to 19, 21
      portability 16
   practice use 31, 45
   programming, steps to 25
      double-checking 30
      standardization 26
         no 29
         yes 28
psychokinetic 43

pendulum, continued
- shapes
  - graphic 17
  - plumbob 16, 19
  - tear drop 19
- suspender .... *See* suspension system
- swing
  - circle 36, 114
  - how 44
  - reading 26
  - who controls 43
  - why 44
- The Ultimate
  - free with book 18, 22
  - research for 18
- training .... *See* programming
- usage
  - as facilitator 13
  - determining sex of egg 41
  - detecting illnesses 81
  - for creativity 12
  - for decision-making 12
  - for predicting 12, 126
    - election outcome 154
  - for profit 64, 126
  - in public 33
  - problem solving 67
  - psychically detect 83
    - for allergies .... *See* allergies
  - when to use 22
  - where to carry 22

pendulum / usage, continued
    versatility 16
    universal standard 26
pendulum-on-a-stick .... *See* bobber
plumbob 16
prana 164
programming 25, 26
    double-checking 30
    no 29
    yes 28
proxy method 85, 106
psychokinesis 43

Q

quadrants 54
    narrowing map size .... *See* map dowsing, how to
questioning
    ask aloud 35
    loopholes 37
    procedures
        being specific 33, 35, 108
    timing, correct 32
    trap 35
        lost earring example 37

R

radon gas .... *See* noxious rays
    testing kit 59, 181
right brain .... *See* inner conscious

S

starting point 71
    graphic 137
success log 61
supernatural 42
suspension system 19, 22, 26
    graphic 27
    how to hold 22, 26
    Psychokinetic Independent 43
swing .... *See* pendulum, swing
symbolic food sample .... *See* proxy method

T

teleradiesthesia. .... *See* map dowsing
Test Button
    for health 72, 84
The American Dowser Quarterly Digest 52, 178
The American Society Of Dowsers, Inc. ....
.... 19, 114, 177, 181
    membership 177
    convention 178
The Dowser's Network (newsletter) 178
Thought Buttons 71
    numbers 1-10 73
        counting procedure 74
    Test Button 72, 84
    percentages 77
        example, The New York Story 77
    X 10 Button 75
    X 100 Button 76
    Y-N Button 71

T continued

thought-energy	43
Time Machine	151
graphic	152, 157
usage	
for future	153
predicting elections	154
past lives	161
present past	158
broken bone example	157
tithing	64

U

Ultimate Pendulum	18, 22
universal standard	26

V

vegan	88
vibes	*See* vibrations
vibrations	
sensing	21, 166

W

wand	16
witness	51, 106

Y

yes, programming	28
Y-N Button	71
Y-rods	*See* forked stick

"YOU *CAN* MOVE THINGS WITH YOUR MIND!!!"

The Energy Wheel® is a device to explore the visible indication of the energy flows around a living body, particularly the hands. It shows you how your mind is able to control these energy flows. In a few simple experiments, you can prove that *the action of this wheel cannot be explained by standard scientific theories today. Even more, anyone can move matter with their mind!* Developed by engineer/healer Gerald Loe, author of *The "Gift" Of Healing*.

THE ENERGY WHEEL®

As featured in the book
MIND MACHINES YOU CAN BUILD

✸ Extremely sensitive ✸ A repeatable psychosynthesis experiment —spins from body (LIFE) energy ✸ For fun and bioenergy research ✸ Study strength and control of your energy field ✸ Control direction of spin by concentration ✸ Some spin it at a distance ✸ Some spin it under glass ✸ Practice with the Energy Wheel® may increase your healing abilities ✸ Fits into pocket or purse ✸ Instructions included

THE ENERGY WHEEL®
PSYCHIC MOTOR
Designed by Gerald Loe

Large 2½ inch Rotor
Complete with plastic
carrying case/stand
**Packaged in book-size box
ISBN 1-56087-008-7
Includes Instructions, $7.95 + $4.00 priority s/h**

Amazing and Wonderful
MIND MACHINES YOU CAN BUILD
by G. Harry Stine

"A collection of how-to instructions for amazing gadgets based on technology we don't understand... yet!"

Travel with one of the greatest scientific minds into the realm of unexplainable technology that works despite the fact that known science says it shouldn't. *Mind Machines You Can Build* is your easy-to-read guide to the forefront of scientific discovery and experimentation. Discover how you can easily build gadgets and gizmos (many with items you now have in your home) to:

- ✳ amplify your psychic power
- ✳ move things with your mind
- ✳ test for your strengths and weaknesses
- ✳ concentrate your energy to accelerate or slow reactions
- ✳ find water, oil, gold, and much more.

Explains metaphysics through understandable science.

Determine For Yourself What Is Hoax And What Is Real!

This is NOT an occult book. It is a foray into inexplicable scientific "laws." With Stine's easy-to-follow instructions, you will be able to construct these devices yourself. Then test them for validity. You can even compare them to Stine's hypothesis on what makes them tick. Break down scientific paradigms. Explore the fascinating fringes of today's scientific mysteries... maybe YOU will encounter the answers to solve these strange occurrences!

"A vastly entertaining book... my friend G. Harry Stine offers here much fun to all of us who refuse to believe that everything worthwhile has by now been discovered."
Gene Roddenberry—creator of STAR TREK

GREAT FOR SCIENCE FAIR PROJECTS

ISBN 1-56087-075-3, Quality Paperback, 208 pages, 34 Illustrations, $13.95 + $4.00 priority s/h

ESP FOR KIDS
HOW TO DEVELOP YOUR CHILD'S PSYCHIC ABILITY
by Dr. Tag Powell and Carol Howell Mills

An unequaled, one-of-a-kind approach to psychic development and children. It is based upon over 25 years of research and psychic development classes taught worldwide. Learn about the unique talents found among children like yours. Written using step-by-step techniques, this book is easy-to-read and includes fun games designed to nurture and improve your child's psychic talents. You will enjoy activities such as spoon bending, psychometry, and psychic healing.

You will :
* enjoy a mind adventure into "The Magic Forest"
* learn to psychically speed grow plants
* discover the healing gift of "laying-on-of-hands"
* master the proven successful techniques of spoon bending
* uncover the mysteries of the past through Past Life Regression
* experience the thrill of Mental Projection and Out-of-Body-Travel
* The "FUN GAMES" are tested and proven psychic development techniques

FOREWORD BY URI GELLER

The authors also explore meditation, dreams, and clairvoyance. With their guidance, you will learn how to tap into you and your child's "hidden" source of creativity, helping to achieve optimum potential! An astounding work.

**ISBN 0-914295-98-5, Quality Trade Paperback
194 pages, Photos, $12.95 + $4.00 priority s/h**

DISCOVER YOUR PERFECT SOUL MATE

by Dr. Judith Powell
the "professional cupid"

Do you want to find your PERFECT MATE? Do you want to enhance the relationship you are already enjoying?

Dr. Judith Powell, an award-winning lecturer, has presented *The Perfect Mate Seminar* around the world. Due to the demand from lonely, frustrated people all over, this cassette tape is now available, containing the highlights of this interesting and matchmaking seminar.

Side A contains a 20-minute seminar on finding your perfect soul mate. You will learn the six steps on programming for a perfect mate, while getting yourself "together" at the same time. You will learn to evaluate your needs and desires; and you will describe specifically what traits and characteristics are important to you in a mate.

Side B guides you to a relaxed level of mind where you can visualize you and your perfect mate together.

SUCCESS COMMENTS

"I finally discovered that the kind of person I was looking for was not really the kind that was best for me. I was setting myself up for failure in love all along. Now it's my turn to really discover my perfect mate."

J.D. (Teacher)

"I found my perfect mate at a seminar, and I programmed my mind for a millionaire! We were married shortly thereafter, and we've been travelling ever since."

M.S. (Real Estate)

"I found my perfect mate — Dr. Tag Powell — in three months! I know that I can help you find your love, too."

Dr. Judith Powell

Contains: Audiocassette, complete handbook and Alphamatic card in book-size box.
ISBN 1-56087-078-8
$12.95 + $4.00 priority s/h.

UP YOUR ENERGY

Your self-power thermostat can easily be turbocharged. Learn to awaken your towering strength, physical ecstasy, and wondrous cheerfulness.

Free yourself from stress and tension. Go for the gold and have the motivation to accomplish all your goals.

Energize your body using your own Inner Power. Turn your mind on for self-empowerment and achieve the great heights you've only previously imagined.

Side A:
CHAKRAS Innergization visualization exercise;

Side B:
Energy subliminals in ocean-type sound.

WINNER of the Benjamin Franklin Award for BEST AUDIOCASSETTE

Contains: Cassette, manual and Alphamatic card in book-size box
by *Dr. Judith Powell*
ISBN 0-914295-35-7 $9.95 + $4.00 s/h

The Ultimate Pendulum

- ✧ The Finest Pendulum Ever Made.
- ✧ Precision-Weighted Brass Instrument.
- ✧ Wind-Tunnel Designed for Least Air Drag.
- ✧ Suspender Specially Designed with 157 Brass Ball-Joint Swivels for Maximum Freedom of Movement.
- ✧ *Bonus...* Black Velvet Carrying Pouch

❑ Yes! I would like to have _____ Ultimate Pendulum(s). *Limit 3 per order.* I enclose $9.95 per Pendule plus $4.00 Priority shipping for full order.
<u>Please make check to Powell Productions.</u>

❑ Please charge to my credit card # _____

_____ Expire Date _____

❑ Please send me your FREE catalog.
❑ I have never used a pendule before.
❑ I have used a pendule before.
❑ I would like to attend a *Taming The Wild Pendulum Seminar*. Please send dates and location.

Name_____

Address_____

City_____ State_____

Zip Code _____

Phone (_____)_____

FAX (_____)_____

Mail to: Powell Productions
P.O. Box 2244, Pinellas Park, FL 34664-2244
FAX 813 3914598 - Phone 813 391-3843

You Asked For Them...
Large Lifetime Laminated Powell MultiMeters

Get Great Results with Your Own Professional Set of Beautiful Four-Color 5 x 8" Powell MultiMeters, Thermolaminated Back to Back to Last Years.

Plus a Bonus of

Four NEW Pocket-Size Multimeters

These handy 4 x 5" Multimeters are also thermo-laminated back-to-back. These are useful pocket charts you can carry with you everywhere you go...

Ready to Work at Your Fingertips at All Times

Your Set Includes: One 4 x 5" and One 5 x 8" Charts of Each of the Following...
1. Basic MultiMeter with Health Band
2. Compatibility Pallette
3. Time Travel Machine
4. New Experimental Spelling Chart

All Eight Powell MultiMeters for only $14.95

Plus $4.00 for Rush Priority Shipping

Name_____

Address_____

City_____ State_____

Zip Code _____

Phone (_____)_____

FAX (_____)_____

Mail to Powell Productions
P.O. Box 2244, Pinellas Park, FL 34664-2244
FAX 813 3914598 - Phone 813 391-3843